MODERN
POULTRY KEEPING

This book sets out the basic principles of modern poultry keeping in a way that combines the vast experience of the British poultry industry with the latest practical ideas from America. It will provide the student of agriculture with a useful textbook and the poultry keeper with a practical work of reference.

TEACH YOURSELF BOOKS

MODERN POULTRY KEEPING

Based on the original work by

C. E. Fermor

Revised and enlarged by

John Portsmouth
N.D.P., N.D.R., N.C.P., F.P.H.

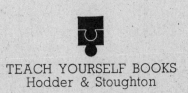

TEACH YOURSELF BOOKS
Hodder & Stoughton

First printed in this form 1965
Second edition 1972
Third impression 1975
Fifth impression 1976
Sixth impression 1977

Copyright ©1965, 1972
Hodder and Stoughton Ltd

All rights reserved. No part of this publication may be reproduced or transmitted in any form or by any means, electronic or mechanical, including photocopy, recording, or any information storage and retrieval system, without permission in writing from the publisher.

ISBN 0 340 16800 5

Printed in Great Britain for
Hodder and Stoughton Paperbacks,
a division of Hodder and Stoughton Ltd,
Mill Road, Dunton Green, Sevenoaks, Kent
(Editorial Office: 47 Bedford Square, London, WC1 3DP)
by Elliott Bros & Yeoman Ltd, Liverpool L24 9JL

Contents

Foreword

The dramatic changes which have occurred in the British poultry industry during the last ten years or so have been reflected in a state of Supply = Demand. With a turnover of £300 000 000 each year the industry is no longer looked upon as the Cinderella of agriculture. This expansion in output has been largely brought about by vast contributions in technical knowledge from the breeding, housing and nutritional fields. Contributions have not been entirely British. The United States of America, never behindhand at converting theory into practice, has contributed much.

With the exception of Chapter 4, dealing with 'The Hen and the Egg', this book has now been completely revised and rewritten, and it sets out the basic principles of modern poultry keeping in a way which combines the vast experience of the British poultry industry with the latest practical ideas from America. Emphasis has been placed not so much on the recognized breeds as on strains and the production of modern hybrids. The now outdated information on free range and semi-intensive housing has been replaced by practical details about battery cages, wire and slatted floors.

Both the traditional principles and the innovations which have been introduced since the book's last revision will be of academic interest and economic importance to those learning or engaged in poultry keeping. *Modern Poultry Keeping* will provide the student of agriculture with an up-to-date textbook and the poultry keeper with a practical work of reference.

List of Illustrations

Chapter 1

Introduction

Tracing the development of the British Poultry Industry from the time of the First World War to the early 1960s, and outlining the different branches of poultry farming currently practised.

The poultry industry today is a complete contrast to what it was before the First World War. All the eggs and poultry were produced by general farmers and smallholders who made little or no attempt to run the poultry unit as a business. The care of the birds was the responsibility of the householder's wife. She regarded any profit arising from her labours as her own money. Nearly all the fowls in the industry at that time were so-called 'dual-purpose' breeds. That is, they were kept for both meat and eggs. All were housed on free range and expected to fend for themselves by finding food around the stack yard. Losses from predators were high and egg production was extremely low. Supplementary feedingstuffs, which were occasionally provided, were not charged against the flock, neither were wages nor other overheads. Keeping hens was not tolerated by the farmer, for he regarded them as more of a nuisance than an asset to his profit and loss sheet. In some quarters this attitude persists even today, although, thankfully, the numbers become fewer each year. The ending of the Great War saw many ex-servicemen enter the industry. Expansion was rapid, with the result that many 'ran before they could walk'. Systems of poultry keeping were employed which had not been thoroughly proven. Breeders concentrated on egg numbers and ignored such important economic characteristics as vitality. The results were disastrous, and a few are even reflected in today's modern poultry industry. One should not, however, forget completely the pioneers of the

1

past, for without such men as Hanson, who proposed the semi-intensive housing system, the industry would probably not have developed at such a rapid rate.

Poultry farming as a business

Poultry farming today is a highly specialized business calling for men with a wide technical knowledge. Today there is no room for the inexperienced, under-capitalized newcomer. Profit margins are small and during the next few years are likely to become even smaller. The get-rich-quick operator is advised to seek his fortunes outside the poultry industry. Today there is no room for him. It is true that ten or more years ago a man could commence with a modest amount of capital and through hard work, and profiting by his successes and learning by his mistakes, succeed. Many of our best poultry breeders started in this way, being content to commence with small flock numbers and build up slowly. It is extremely doubtful that he would succeed today, for capital requirements are high and the pattern in the industry has changed completely.

Poultry farming need no longer be a seven-day-a-week occupation, although it still is on many small-sized units. New equipment and modern ideas, together with work study organizations, have resulted in fewer working hours for many of the large commercial unit employees. Obviously, though, where much automation is employed there is always the risk of mechanical failure and some labour, skilled in light engineering, is employed. Thus, whilst the employees' working week may be reduced, management staff often shoulder heavier weekend responsibilities.

One-man units are, for obvious reasons, tied to the poultry farm more than the large-scale operation, but even so with careful work planning the owner need not sacrifice his whole life to work.

Growth of the industry

To appreciate the extent to which the poultry industry grew between the opening of the Second World War and

1960 it is only necessary to point out that, according to the Ministry of Agriculture's Returns for England and Wales for the period 1937–8, the estimated value of output in respect of poultry and eggs was £22 600 000. For 1959–60 the estimated returns were £300 000 000.

The British poultry industry accounts for one-fifth of the total value of our agricultural industry, which is worth approximately £1 500 000 000 per annum.

In the expansion that has taken place the different facets of the poultry industry have each assumed greater importance in its overall worth.

In every section the country is virtually self-supporting. Indeed, at the present time, there is a major threat of overproduction in both eggs and table poultry which has led to a reduction in the poultry farmer's profit margins. Whilst there are large egg farms, breeding has become a very specialized business, and we now have large hybrid breeding establishments and hatcheries. Many are British-owned but a number are American-supported.

The production of table poultry is no longer the side line of the egg producer but a highly specialized aspect of the poultry industry. There are scientifically run broiler and turkey plants. Packing stations, either privately- or producer-owned, market both eggs and poultry in return for a small service charge.

The newcomer

What branch would a newcomer to the industry wish to take up? A brief outline of each will help him to form an opinion.

To many the appeal of pedigree breeding will be great, but this side of the industry requires much skill and experience. The work of the pedigree breeder today is to supply the larger breeding organizations with pure-bred stock from a closed flock. In the past the breeder was the backbone of the industry, supplying day-old's, growing stock and adult breeders to commercial breeders. The commercial breeder who was involved in cross-mating turned to him for a change of blood in the form of a male bird to head his breeding pen. The birds were trap-nested and meticulously recorded seven days a week. He could supply information to purchasers of his stock which gave details of the dams' records and records of the sire's dam also. Today this work is carried out by large chick-producing organizations who employ several geneticists and statisticians to prepare and evaluate the work.

Hatching work is also carried out by these large organizations employing scientifically trained incubationists. Day-old pullet chicks are hatched by the thousands each week in large 'walk in' incubators in which the temperature and humidity are automatically controlled. This work, along with breeding, is obviously the province of one who has high technical qualifications and has been engaged in the poultry industry for a number of years.

The egg farmer

Most people when they are starting in poultry keeping turn to egg production, because they feel it is one side of the industry which requires least experience.

Unfortunately, the production of table eggs involves more than feeding and egg collection. Knowledge of housing, feeding, general management and disease control are essential.

One must know how to handle birds, to see whether they are fat or out of condition, and what steps to take to apply a remedy. The poultryman should be able to distinguish the layer which is producing economically from the non-producer losing money.

If he is purchasing day-old chicks or growing birds, he will require knowledge of brooding and rearing. It should never be forgotten that poor rearing can obviously affect the birds' laying potentials. The reader will quickly see that the egg producer has plenty to do, without necessarily taking on breeding or incubation.

Table poultry

The activities of the table poultry plant will, as its name implies, be confined to the production of poultry for the table. This is an all-the-year-round job of fattening birds for human consumption.

Table poultry production may involve several separate operations, i.e. broilers, capons, roasting fowls, turkeys, ducks and geese.

The main advantage of table production is that the capital involved is turned over more frequently. The initial capital outlay, when compared to egg production, need not be so great, although this will depend a great deal upon the class of stock reared.

The work consists wholly of rearing birds that will carry the maximum amount of flesh in the shortest possible time and at the least cost.

Most of the table birds produced in any quantity are marketed through a packing station, which provides the service of processing and marketing for a small charge. Only small table production units kill and prepare their own produce, as the marketing side has become extremely competitive.

The broiler industry is extremely specialized. Profit margins are small and likely to remain so. Therefore, unless large numbers are reared for killing, the return will not be great. Broilers are marketed at eight to ten weeks of age. They are birds of both sexes weighing between $2\frac{3}{4}$–$4\frac{1}{2}$ lb liveweight.

The smallholder and backyarder

The smallholder is usually a one-man unit employing family labour when required. He may not rely solely on poultry for a living but have other farmstock also. He is most likely to be interested in small-scale egg production and marketing his own eggs over the farm gate. Small-scale capon and turkey production may also be considered.

The backyarder or domestic poultry keeper is concerned only with supplying his family and possibly a few friends with fresh eggs. Flock size will therefore be small, and the majority will purchase ready-to-lay pullets and not consider raising their own. A few D.P.K's hatch their own chicks under broody hens. This obviously reduces costs, but additional

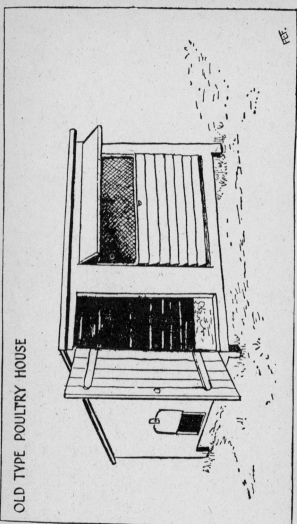

OLD TYPE POULTRY HOUSE

Fig. 1 Usually associated now with Backyard Poultry keeping.

knowledge of chick rearing and broody hen management is required.

The general farm

In this day of specialization it is not surprising to learn that few of all the eggs produced in Great Britain are supplied by general farmers with flock sizes of less than one thousand birds. Many general farmers, because of the availability of farm buildings and surplus cereal grains, are able to supplement usefully the general farm income. Many take advantage of stubble fields by ranging fattening cockerels over them during the autumn. In this way food costs can be reduced, although in fox-prone areas losses could well be high.

General farmers are quickly realizing the important part that poultry can play in the farm economy. Many are not only using existing farm buildings but also erecting specially designed poultry houses. The value of poultry manure to their land is also important.

The poultry holding

Nothing has been said about the poultry holding itself. Ten or more years ago, when large numbers of poultry were housed on free range, the type of soil, position and natural wind breaks were points for serious consideration. Today, however, less importance is attached to these points because stock is now housed intensively. Houses which control both temperature and light are called controlled environment houses. Their use virtually eliminates the necessity for siting buildings according to the sun and prevailing winds.

Where free range rearing is still practised, thought must obviously be given to soil type and drainage, for these factors directly affect the stocking density of the fields.

Hard roads are essential, as also are main water and electricity. Without these amenities labour costs and relations may be embarrassing.

Chapter 2

Systems of Poultry Keeping

Free range, semi-intensive and intensive systems.
Their advantages and disadvantages.

After deciding what branch of poultry keeping to adopt, it is necessary to have some idea of what system to use.

There are several ways of keeping poultry. Some have been in use for a good many years, others are comparatively new, but all have by now had a good trial on both large and small commercial holdings.

When dealing with the non-specialist poultry farmer it is difficult to say that there is any best way, for such a lot depends on circumstances. For example, many general farmers have adapted empty barns and outbuildings for running poultry on the deep litter system. The specialist poultry farmer, on the other hand, uses the battery cage system which, although more expensive to capitalize than deep litter, gives a greater return on the capital invested.

No system should be condemned as a system, but rather condemnation should come from the use of any system, however good in itself, in circumstances which do not justify it.

The various methods practised may be grouped as follows: (i) extensive or free range, (ii) semi-intensive and (iii) intensive.

Free range

The free range system is little used by the commercial poultry farmer. Its use is confined to small poultry keepers, a few general farmers and D.P.K.'s.

The system allows the birds to range at will over an almost

unlimited area of ground, although as a yardstick 300 adult birds to the acre is generally followed. Housing is simple and incorporates a sleeping and laying area. Occasionally, field houses are used which have feeding troughs attached to the inside so that the birds can feed under shelter during bad weather.

THE FOLD

Fig. 2 Popular type of housing for layers.

Production is generally poor, comparatively speaking, because the birds are exposed to all weather conditions. Artificial lighting is difficult to provide, and because of this egg production is poor during the autumn and winter months. It is highest during the spring and summer.

Growing poultry may be profitably reared on free range during the spring, summer and autumn, but winter rearing is extremely difficult and losses may be high.

Several years ago the folding system was commonly used and recommended. Birds were confined to a wooden house and wire netting run. The unit measured some 24 feet in length by 4 feet wide. A section at one end measuring

4 feet by 4 feet was used for sleeping. Nest boxes were attached to this section. Each unit housed twenty-five adult laying birds or fifty growing stock. They were moved daily to prevent the ground becoming 'poached'.

Besides being a protection against foxes the fold had other advantages. Once the birds had become accustomed to the sleeping compartment there was no necessity to let them out or shut them up in the mornings and evenings respectively. As it was possible to provide a group of field units with artificial lights and as protection could be given in bad weather, egg production was higher compared with free range. However, the need for moving them daily was expensive in terms of labour. Regular moving also caused wear and tear and higher depreciation charges on the equipment.

From a general farmer's point of view, the manure voided was valuable because it was very evenly distributed over the ground.

The semi-intensive system

As the name suggests, this system is a compromise between the free range and intensive systems.

This system is almost completely confined to breeding farms where the scale of operation is not large. It is rarely used today on commercial holdings because, for a small additional capital outlay, laying birds may be housed completely intensively and thereby be more easily controlled.

The semi-intensive system involves a static house, to which is attached two grass runs. One run is used for about six months, rested for six months while the other is used, and then the cycle is repeated *ad infinitum*. Sometimes the semi-intensive system involves one run, which is rested periodically, ploughed up and then resown with grass.

As has already been mentioned, this system is mainly a compromise between free range and complete intensivism. Provided precautions are taken against overcrowding, the birds may be completely confined during bad weather, to the benefit of both the birds and the stockman. The disadvantages of the static or permanent semi-intensive house is the danger of the ground around the house becoming stale.

SEMI-INTENSIVE HOUSE

Fig. 3   Birds can be kept in during inclement weather.

When this occurs the disease risks are at once increased. Precautions may be taken to avoid this by removing the top spit of stale soil and replacing it with fresh soil.

Intensive systems

Intensive systems are used because they save labour, permit almost complete control of the birds' environment, reduce production costs and, most important of all, increase production efficiency.

As will be realized, it is a system that can be practised with advantage on poultry farms which are situated near towns and where land prices are high or in short supply.

Whilst the cost of housing is more expensive compared to the other systems discussed, it is the most economic.

Experience of poultry keeping is absolutely essential with this system, for overcrowding and errors in feeding can lead to cannibalistic activities amongst the birds.

Housing

Housing and appliances represent a considerable proportion of the capital invested in the poultry farm.

Much thought must therefore be given to the selection of houses and to their construction, because initial mistakes are easily made and costly to rectify.

Poultry buildings have one of three types of roof: (a) lean type, (b) span roof and (c) uneven span.

(a) Lean type are easy to construct and are most suitable for small buildings. In large lean-to houses natural ventilation is not easy to regulate and mechanical ventilation may be needed. (b) The span roof has a ridge which is equal distance from each wall. Ventilation need not be difficult as outlets can be incorporated along the ridge. (c) The uneven span has one side a greater distance from the walls than the other.

Ventilation

Adequate ventilation is absolutely necessary with intensively-housed poultry. In the absence of good ventilation the incidence of respiratory infections is increased.

In hot weather a good ventilation system keeps the birds comfortable. The siting of both inlets and outlets in intensive poultry houses is of vital importance. Air must not be brought directly into the building but should be baffled to prevent draughts.

Mechanically controlled ventilation is now frequently used, especially in intensive houses, where the light and ventilation are accurately regulated.

Ventilation can be based on the following yardstick.

Extract 1 cubic foot of air per minute per bird for winter ventilation. During the summer months when air temperatures are higher the extraction rate should be increased to between 6 and 10 cubic feet per bird. One 24-inch extractor fan will extract approximately 5500 cubic feet of air per minute and in order to balance the ventilation 4 to 5 square feet of inlet area should be allowed for each 1000 cubic feet per minute. On this basis of 10 ft^3/min per bird in summer a useful ventilation yardstick is one 24-inch fan for each 550 laying birds.

Insulation

To obtain satisfactory air movement the house must be warm. Insulation maintains house temperature during cold weather. It also keeps the house cool in warm weather. Materials such as fibre glass, wood wool and insulation board are good insulators, whilst asbestos, glass and corrugated iron are poor insulators.

The roof is the most important part of the house to insulate. Good insulation, whilst adding to the capital cost, can save considerable amounts of money in the form of lower food consumption and more even production.

Deep litter system

Apart from the increasing popularity of the battery cage system of poultry keeping, deep litter housing is still extremely popular and used by many poultry keepers. The litter must be dry and crumbly. It should not be damp and firm. Deep litter is made by adding 4 inches of wood shavings or peat moss to the floor of the poultry house. The poultry manure decomposes the shavings, and by adding a further 1 inch of new litter each month a depth of 9 – 10 inches is reached within six months of starting the litter.

Each bird should be allowed 3 square feet of floor space, 5 inches of feeding space and 1 inch of water space. Allow 1 square foot of nest box area for each four birds. If perches are used, allocate 9 inches per bird.

The part-wire or part-slatted floor deep litter house is more commonly used today than deep litter alone. This is because more birds can be housed at less cost and the litter can be kept much drier, as a large proportion of the birds' droppings fall under the wire or slatted floor area.

Allow each bird 2 square feet of total floor area. As perches are unnecessary with this system, 9 square inches of wire or slatted floor should be allocated to each bird, leaving 1¼ square feet of littered area per bird.

The battery method

Laying cages afford one of the most efficient means of commercial egg production. The main advantage of using laying cages is a high rate of production. This is because, in comparison with other systems of management, interference by other birds is considerably reduced. Culling and handling is also made simpler. Its main disadvantage is in the inflexibility of the system, i.e. battery cages are suitable only for poultry.

A single bird-laying cage has the following dimensions: width 9 inches to 12 inches; depth 18 inches; height 18 inches, sloping to 14 inches at the back. Eggs are laid on the wire floor and roll into a cradle, which extends about 6 inches beyond the cage front.

Cages can be designed for more than one bird. With the exception of the width, the other dimensions remain constant, irrespective of the number of birds.

The following figures will give the reader an idea of the different cage sizes used, together with the number of birds housed in each size.

Inches	Number of birds
9–10	1 bird
12–13	2 small or 1 large
14–15	3 small or 2 large
17–18	4 small or 3 large
21	5 small or 4 large
31–32	7 small or 6 large
42	9–10 small or 8 large

Battery cages can be made fully automatic, including egg collection. Feeding is by motor-driven food troughs which travel up and down the front of the cage filling the open troughs. Water can be provided by open gutter troughs supplied from small header tanks situated on the top of each block of cages. These are refilled from the main tank, usually situated in the roof of the building. So-called nipple valves are also used to supply water to the birds. Drinking is achieved by the bird depressing the moveable nipple which allows a stream of water to flow.

Manure removal is automatic on many battery cage units. Metal scrapers pass beneath the birds, pushing their droppings to one end where they can be manually or automatically removed. Manual removal involves the use of a shovel and wheelbarrow. Automatic egg collection is the exception rather than the rule. It involves the transfer of eggs, as they are laid, to the end of the unit, where they may be man-handled or automatically conveyed to a collecting area.

Deep pit houses

In order to reduce the amount of labour necessary for manure disposal, some battery cage units are sited over 6 feet deep pits into which the droppings collect. Removal of the manure may be annual or biannual depending on the depth of the pit.

Flat duck battery cage system

Instead of placing three blocks of cages one above the other, the flat duck system spreads horizontally. Thus, just one tier of cages is used spread across the floor area. Gangways for normal working may space the cage blocks. Cat-walks placed on top of the cages may also be used and gangways eliminated.

Chapter 3

Breeds and Breeding

A brief account of the various breeds in common use.
The use of cross breeds. Heredity.
Breeding for economic characteristics. Modern hybrid production. Sex linkage. Management of breeding stock.

In selecting a pure breed, cross-bred or hybrid, the poultry farmer should first be guided by its overall performance at other locations, particularly at Random Sample Trials, in the case of the commercial producer. The backyarder or fancier will, to a great extent, be influenced by colour, type and other outward characteristics which appeal to him,

The commercial egg and meat producer should not be guided by points which influence the fancier.

The businessman will look to the popular breeds and crosses, for evidence of popularity will be found in ability to prove a money-maker. From the purely commercial aspect, the poultry farmer will concentrate on purchasing hybrids renowned for their good performance at trials. As indicated in the introduction, breeding calls for much skill and experience, and only those knowingly capable of following this aspect would purchase or consider the pure breeds as we know them today.

The table set out in this chapter deals with a number of breeds which have been used in the past as the foundation types of a modern poultry industry. Today the emphasis is away from pure bred stock and directed towards hybrids, of which, although some, for all intents and purposes, are produced from only one breed, the main are the product of two or more separate pure types being intercrossed.

Of the breeds named certain of them are egg producers in their own right, and in these cases the cockerels do not prove a profitable proposition if fattened for the table. The general purpose breeds still exist today, although it is accepted that no one breed type is capable of high egg laying and producing a first-quality carcass on the male side—at least, not to obtain maximum profit.

Table poultry farmers, e.g. broiler producers, require types with excellent table qualities, that mature quickly and convert food into meat efficiently.

Name of breed	Country of origin	Comb	Colour of flesh and skin	Colour of egg-shell	Colour of legs	Remarks
Rhode Island Red	America	Single	Yellow	Tinted or Brown	Yellow	General Purpose Breed
Sussex	Britain	,,	White	Tinted	White	,,
Leghorn	Italy	,,	Yellow	White	Yellow	Laying Breed
Wyandotte	America	Rose	,,	Tinted	,,	General Purpose Breed
Plymouth Rock	America	Single	,,	,,	,,	,,
Buff Rock	America	,,	,,	,,	,,	,,
Ancona	Italy	,,	.,	White	,,	Laying Breed
New Hampshire Red	America	Single		Tinted	,,	General Purpose Breed
North Holland Blue	Dutch	,,		,,	White	,,
Maran	France	,,		Brown	,,	,,
Dorking	Britain	Single (except White)	White	White	,,	Table Breed
Game (Indian)	Britain	Pea	Yellow	Tinted	Yellow	,,
Game	Britain	Single	White	,,	White	,,
Orpington	Britain	,,	,,	,,	,,	General Purpose or Table
Faverolle	France	,,	,,	,,	,,	,,

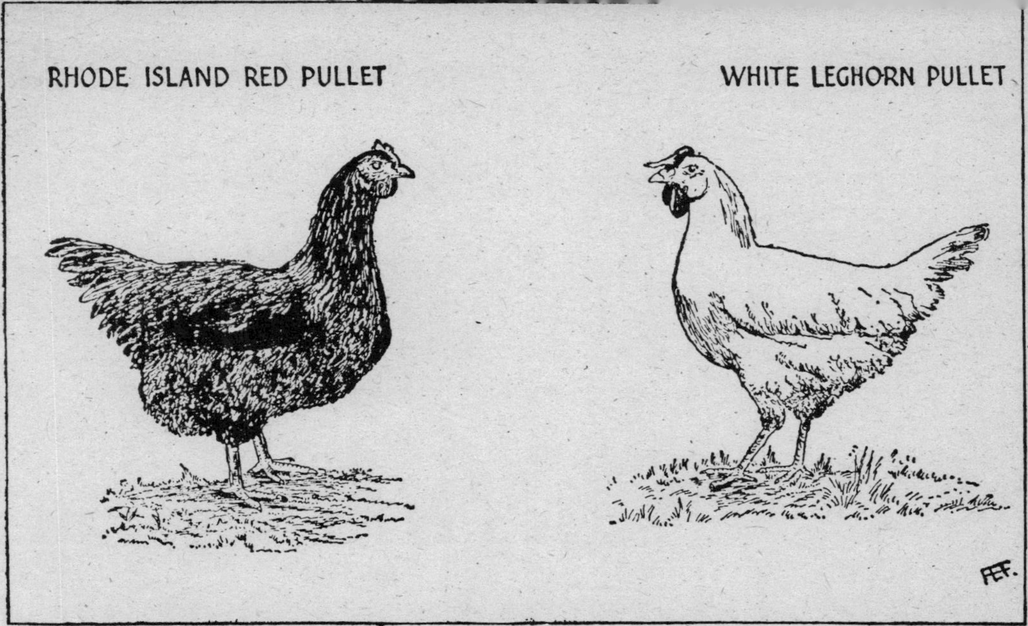

RHODE ISLAND RED PULLET

WHITE LEGHORN PULLET

Fig. 4

One of the most popular breeds.

Layer of large white eggs.

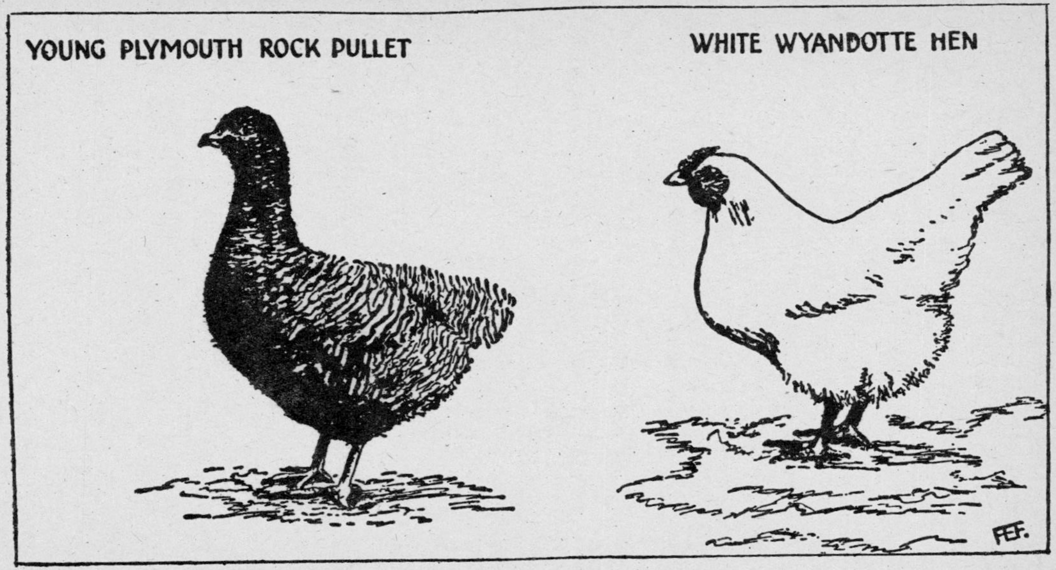

YOUNG PLYMOUTH ROCK PULLET

WHITE WYANDOTTE HEN

Fig. 5

A good laying breed.

A good layer of tinted eggs.

It is probably true to say that, although only a handful of those breeds mentioned are found on a few commercial units today, three-quarters of them are used in the production of hybrids for the egg and table poultry industries. It is important to remember that a breed itself is relatively unimportant. What matters, of course, is the strain within a given breed.

Bearing the last sentence in mind, the following notes describe a few of the pure breeds used today in the production of commercial crosses and hybrids.

Rhode Island Red. It is a rich chestnut colour and has proved a very good layer in various systems of management. As a table bird its qualities are limiting. It has a high breast bone and yellow skin, which detracts from its value. The breed has been used extensively in crossing with Sussex hens (a sex-linked cross). Sex linkage is explained later in this chapter.

Sussex. Now used widely for both egg and table production, although prior to the last decade the breed was regarded purely as a table bird. The Light Sussex is the most popular variety. The plumage is basically white, with black ticking on the neck. It has a black tail.

Leghorn. This breed belongs to the non-sitting (non broody) light breeds. The best known varieties are the white, black and brown, of which the white is the most popular today. It is used extensively in hybrid production for crossing strains within the breed and also for crossing with other breeds, notably the Rhode Island Red. It is a most prolific layer of white eggs. Its table qualities are poor.

The White Rock. This bird is a variety of the Plymouth Rock. Both are used for crossing purposes in the production of broilers. Many commercial broilers on the market today have White Rock in their make-up.

New Hampshire Red. Similar to the Rhode Island Red but slightly lighter in colour. This breed was developed primarily for its table qualities and is now used extensively in the production of commercial broilers. It is much faster maturing and has better feathering than the Rhode Island Red.

INDIAN GAME COCKEREL

DARK GREY OLD ENGLISH
GAME COCK

FEF.

FEF.

Fig. 6 Possess good table qualities.

Crossed with the Light Sussex female it gives 100 per cent sex linkage.

Game Fowls. The two chief game fowls kept in this country are the Indian or Cornish Game and the Old English Game.

As pure breeds, both are slow maturing and poor layers, but when crossed with the more prolific breeds egg production is improved in the progeny. The Indian Game carries an abundance of breast meat, but because of extreme breast width it gives poor fertility. The Indian Game has been used quite considerably in the broiler industry.

The Old English Game is an excellent table bird particularly when crossed with the Game breed mentioned above.

French table breeds. Only the Faverolle combines large size with early maturity, good egg yield and fine white flesh. Used in crossing with the Sussex, a very fine table bird is produced.

La Bresse is quite popular, the white variety being most common. This breed has fineness of bone, good breast conformation and excellent white flesh.

Both these breeds were suitable for table poultry production a generation ago. Today, in comparison with modern strains and crosses, they grow too slowly for economic requirements.

Cross-breeds

The question is sometimes raised, 'Are cross-breeds better than pures?' From a commercial standpoint, the crossing of two different breeds results in lower mortality, faster growth and sometimes greater egg production. This does not always hold true, for much depends on whether the two breeds complement each other in terms of genetic make-up.

It is unwise to breed from cross-breeds. They rarely reproduce true to type.

Some commonly used crosses are:

BLACK RED OLD ENGLISH
GAME COCKEREL

SILVER GREY DORKING COCKEREL

Fig. 7 Well-known Table Birds.

For egg production:
> White, black or brown Leghorn crossed with a Rhode
> Island Red.
> Light Sussex crossed with a Rhode Island Red.
> White Leghorn crossed with a Light Sussex.

For table poultry production:
> Indian Game crossed with a Light Sussex.
> New Hampshire Red crossed with a Light Sussex.
> North Holland Blue crossed with a Light Sussex.

Heredity

In order to understand the mechanics of sex linkage, and
also to assist the reader to digest the section dealing with
hybrid production, a brief account of inheritance is provided.

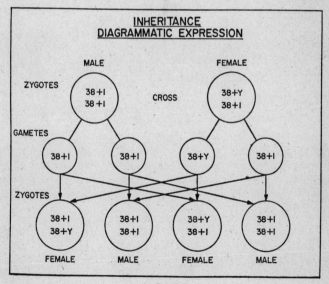

Fig. 8

The reproduction cells of the male and female chicken are called 'gametes'. Thus both sperm and ova are gametes. After fertilization the gametes fuse to form a 'zygote'.

Each gamete contains what are called 'chromosomes'. It is these which are transmitted from parent to offspring. The domestic fowl has thirty-nine pairs of chromosomes which are of two types, i.e. sex chromosomes and autosomes. The number of autosomes remains constant for each sex, but the sex chromosomes vary in respect of sex. The male bird has two sex chromosomes, the female only one.

In poultry it is the female that decides the sex of the offspring because she has only one sex chromosome, plus a so-called 'Y' chromosome. A simple diagram will help clarify this point (see Figure 8).

The reader should understand that the chromosomes are the bearers of the determiners of the hereditary characteristics, because there are only comparatively few chromosomes and literally hundreds of characteristics. Each chromosome is responsible for the development of many characteristics.

With economic characters the environment in which the bird lives can play a major role in marking or potentiating the effects of inheritance. This is why such a wide variation in performance exists amongst a genetically uniform flock of birds.

Breeding for economic characteristics

Commercially, of course, poultry are bred for meat and eggs. The efficiency with which these utility operations can be carried out depends, as mentioned above, mainly on genetic factors. Performance is, however, influenced by environmental factors such as housing, nutrition and management.

In egg production there are five main characteristics. These are: (a) number of eggs, (b) sexual maturity, (c) rate of lay, (d) broodiness and (e) length of lay.

All these factors are important where the selection of breeding stock is concerned.

Production of hybrid chicken

Hybrids are produced by crossing two strains of parent birds which, though they differ genetically, complement each other in respect of their genetic make-up. Cross-mating is mentioned, for whilst this is true it does not necessarily infer crossing two completely different breeds. Not every cross-bred pullet possesses hybrid vigour. Hybrid chickens are not only produced with laying birds but table poultry also.

Hybrid poultry are produced in different ways, but each is based on parent strains that have been selected on their ability to complement each other. This is called 'nicking'. It is usual to test a number of different strains so that the best can be used for producing the commercially marketable end-product. Some hybrid producing schemes involve close inbreeding, which initially causes a lowering of quality but makes the 'line' or 'strain' more uniform in respect of the characters required in the breeding programme.

With hybrid table chickens a similar procedure is followed, although the characteristics will obviously differ.

Assessing the comparative performance between hybrids, cross-breeds and pure breeds is not simple. By definition the modern hybrid should be superior, but this is not always the case. However, hybrid breeding does usually result in a group of birds being more uniform compared with the traditional breeds or crosses. This fact alone has led to their great popularity in recent years.

Sex linkage

By making certain matings, characteristics emerge in the progeny whereby the pullets can be distinguished from the cockerel chicks at day old. This is known as sex linkage. It has already been seen that the female is the determiner of sex in poultry. The female also carries certain characteristics on her sex chromosome which are dominant to those carried by the male. For example, the cross between a Rhode Island Red cockerel and a Light Sussex hen gives rise to brown pullets and silver cockerels at hatching. This is known as the 'gold cross silver' sex linkage. It is the silver gene of the

SEX LINKAGE

LIGHT SUSSEX HEN

RHODE ISLAND RED COCKEREL

FEMALE CHICK

MALE CHICK

Fig. 9

female which is dominant. The reverse cross, namely Light Sussex cockerel with a Rhode Island Red hen, gives rise, as we should expect, to all-silver chicks. This criss-cross inheritance is not only confined to gold and silver feathered birds but can also be produced by crosses involving slow and fast feathering breeds (slow feathering is dominant to fast feathering, and barring to non-barring) (*barring is dominant to non-barring*).

The practical importance of sex linkage is well known to poultrymen today and in theory quite a number of characteristics may be employed to help us sex chicks, but in practice only a few are really suitable.

The following breeds may be used in producing the gold and silver sexed-linked cross:

Rhode Island Red (male) crossed with a Light Sussex (female) (see Figure 9).

Buff Plymouth Rock (male) crossed with a Light Sussex (female).

Brown Leghorn (male) crossed with a Light Sussex (female).

Indian Game (male) crossed with a Light Sussex (female).

New Hampshire Red (male) crossed with a Light Sussex (female).

Management of breeding stock

Physical selection of breeding stock.

Physical handling of breeding stock is an important aspect of selection, for although complex breeding systems and statistical analysis are now commonly used they tell the geneticist nothing whatsoever about the bird's physical conditions.

Handling breeding stock eliminates birds which are physically incapable (no matter how good their records may appear) of making suitable breeding material.

Selection should be based on handling qualities of the stock at regular intervals during rearing and at point of lay. The breeder must be prepared to kill off or otherwise dispose of every breeding bird that shows loss of vigour. Failure to recognize this can reduce efficiency in the flock.

The bird's eye is described as a mirror of vitality. This is true, because the eye reveals the bird's state of health. Structural eye defects should always be regarded with suspicion. The bird's stance should be good and the body should handle firm, not loose like an empty paper bag. Cockerels, too, must be fit and healthy. The birds that by their appearance, behaviour and handling qualities have every evidence of possessing an abundance of vigour should give a good account of themselves in the breeding pen. There can, of course, be no assurance on this point.

The most conducive evidence of vigour is the ability to transmit economic characteristics, to maintain a satisfactory level of production and to produce healthy progeny.

The age for mating breeding stock

Breeding stock must be physically mature before they are used in the breeding pens. This means that for the Light Mediterranean breeds they should be five to six months old and for heavier breeds six to seven months old. Immature breeding stock may breakdown midway in the season. Cockerels may be one month older than pullets, but this is not essential.

A few years ago only yearling or second-year hens were used for breeding, because it was reasoned that if they could live and lay well for one year they would transmit their valuable characteristics to their progeny. In many respects the reasoning is extremely sound, although today pullets or first-year birds are freely bred from, with no detriment to the progeny.

It is well to make up the breeding pens in good time before the eggs are required for incubation. Allow at least four weeks, and preferably six, before collecting hatching eggs. Some of the eggs will be fertile in about ten to fourteen days after mating.

After the male has been removed from the breeding flock, eggs will remain reasonably fertile for a similar period. Eggs should be collected at least three times each day and kept at a temperature of about 55°F.

The time of year for hatching

Today, in commercial pullet chick production, eggs are incubated in every month of the year, although the bulk are hatched in the six months November to March inclusive. Small poultry keepers purchase most of their chicks in the spring months, whilst the larger poultry farmers purchase their chicks all the year round. This is because the larger commercial concerns carefully follow the short-term price trends arising from the established free market for eggs, whereas smaller concerns may be more influenced by season and weather conditions.

In table poultry production and, in particular, broiler production hatching is spread out evenly all the year round, although the demand may vary according to the seasons.

Mating systems

With pedigree work one male may be mated to twelve or fifteen females, and the birds are trap-nested to identify the eggs laid by each hen. These are then carefully followed through incubation and after hatching to provide information on the breeding and economic potential of each trap-nested hen. When a valuable cockerel is required for mating with as many hens as possible, a cyclic breeding system may be used. This involves possibly three breeding pens of twelve females each. The cockerel is left in each pen for two or three days and then transferred to the next. After two to three days in this pen he is moved to the third pen. The system is then repeated.

Flock mating involves mating eight or ten males with each 100 females in the same pen. Individual identification of eggs laid by certain females is possible, but there is no knowledge of which cockerel mated with which hen and thus pedigree breeding is impossible.

Flock mating is used by breeding stock multipliers whose job it is to produce commercial hatching eggs for the large chick-hatching organizations.

Note: Figure 28, page 106, shows the method used for identifying the progeny of particular matings.

Chapter 4

The Hen and the Egg

*Being a description of the external characteristics of
the bird together with details of the internal organs,
including the digestive system and the reproductive
system. Details of the composition of the egg and
blemishes which occasionally arise.*

Origin

It is generally accepted that our domestic fowl originated
from the jungle fowl of India. By changes in condition,
feeding, selection and mating in the past, increase in size
has been attained together with greater fecundity. The
jungle fowl would lay only about twenty-five eggs in a
breeding season and would weigh between 3 and 4 pounds.
The present-day layer will scale between $3\frac{1}{2}$ and $5\frac{1}{2}$ pounds
and may lay up to 250–300 eggs in a year. She would, in
fact, be rejected from the flock if she did not do better than
produce 200 eggs annually.

Whilst no attempt can be made in a general book on
poultry to go fully into the question of the science of the
life (the biology) of the fowl, it is felt that some knowledge
of the make-up of the hen and the egg will be acceptable
and advantageous to the poultryman.

Our illustration here (Figure 10) is not, as a matter of fact,
of a hen at all but of a male bird. It has been used here to
show one or two external characteristics (spurs for instance)
that the female does not possess. The illustration practically
speaks for itself. Notice the feathers, and be able to distinguish between primary and secondary flight feathers. Look
at the sickles, the saddle and the hackle feathers. It should
be noted that in most breeds (the Campine breed is an exception) the males have feathers over certain parts of the

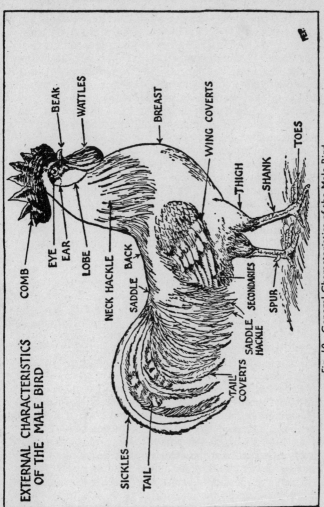

EXTERNAL CHARACTERISTICS OF THE MALE BIRD

COMB

BEAK

WATTLES

EYE
EAR
LOBE

BREAST

NECK HACKLE

WING COVERTS

SADDLE BACK

THIGH

SECONDARIES

SHANK

SADDLE HACKLE

SPUR

TOES

TAIL COVERTS

SICKLES

TAIL

Fig. 10   General Characteristics of the Male Bird.

body that are longer and more pointed than feathers in the same parts of the body of the female. Such features in animals (including birds) are termed 'secondary sexual characters'. In fact, secondary sexual characters are features of the body which though not themselves part of the reproductive system of the male or female are nevertheless typical of that sex and are sufficient in themselves to distinguish it. The hackle, saddle and sickles are 'secondary sexual' feathers. The comb and wattles, which are secondary sexual features, are outgrowths from the skin.

The openings of the nostrils (external nares) should be noted above the tip of the beak. The ears can be found by separating the feathers behind the eyes. Notice the scales on the legs and the shanks, and the spur, peculiar to the male bird. The latter, which is of a horny character, is an outgrowth of the bone.

All poultrymen should make themselves familiar with the make-up of the fowl, if only as a matter of interest.

Internal structure

It is useful, however, to help one recognize changes in any of the organs due to disease. When a bird dies, if it is not wanted for post-mortem examination, it should be dissected and examined, the appearance and structure of the various organs being noted. To do this is a very useful and instructional practice, if the time can be made available. The knowledge gained can be of great service.

In order to open a bird for examination, pluck the feathers from the breast—in fact, from the whole of that side—and place it on a board, breast upwards, and fasten it down, say with nails. If the skin is lifted, one should be able with a sharp pair of scissors to cut through it down from the neck of the cloaca or vent. Turn the skin right back and carefully cut away the flesh from the sides of the sternum or breast bone. Lift the sternum and, after detaching the underlying organs with some blunted instrument, remove it. If the lining or membrane is then carefully removed, the organs will be seen in position.

The heart, conical in shape, will be found in the middle of

the body. The pointed end faces backwards. The lungs, red in colour, lie between the ribs behind the heart. If the skin of the neck is removed, it will be easy to see the windpipe up to where it opens into the back of the mouth.

The liver can be distinguished by its reddish brown colour.

The alimentary or food canal can be traced from the mouth to the cloaca or vent. The part of the food canal which begins at the back of the mouth is known as the oesophagus or gullet. It passes through the neck, opening into the stomach or proventriculus. The crop will be found as an enlarged portion of the oesophagus at the base of the neck.

The proventriculus or stomach comes between the oesophagus and the gizzard. It will be identified as being larger than the oesophagus and is reddish in colour. The gizzard is the fowl's grinding machine. It has thick muscular walls, as will be seen if it is cut open. The food is ground between the horny lining of the gizzard and the grit consumed by the bird, which will be found in the gizzard when it is opened.

The small intestine is a long tube extending from the gizzard to the caeca. These latter are a pair of blind-ended tubes which mark the beginning of the large intestine. The large intestine extends from the caeca to what is known as the cloaca. This is a cavity, into which the large intestine opens.

In the first loop formed by the small intestine, which is known as the duodenum, is a long gland, the pancreas. This is a compact pinkish organ. The small greenish body against the liver is the gall bladder. This stores some of the bile which helps in the digestion of some of the fats contained in the food. A duct leads from the gall bladder to the duodenum.

Digestion

If the bird's mouth is examined, it will be seen that it has no teeth. Any food picked up in the beak is carried into the gullet by the tongue, and from there into the crop. This crop serves the purposes of a food store, and it is here also that grain undergoes a softening process from saliva coming from the glands of the mouth. When food passes from the crop into the proventriculus or stomach, gastric juices are released, which act on the food. In the gizzard the food is gradually reduced to a fine paste by grinding, the gizzard making up for the lack of teeth in the bird. As a paste the food passes into the small intestine, where the process of digestion begun by the action of the saliva and the gastric juices is continued by bile and pancreatic juices. Bile is stored in the gall bladder until required in the small intestine. The effect of all these juices is to convert the proteins, fats, carbohydrates and mineral salts into forms which can be absorbed by the small intestine. During their passage through the intestine, the soluble contents of the food so broken down are absorbed through the walls of the intestine, the material which reaches the end of the food canal being the indigestible food residues.

In the male bird the essential paired organs of reproduction are the testes or testicles, which in a mature bird will easily be seen, one on either side of the backbone, just above the kidneys and below the lungs; the vas deferens or sperm duct is a tube leading from each testicle to the cloaca.

In the female bird, the organs of reproduction consist of an ovary and an oviduct. The ovary produces the yolk of the egg or ovum, and the oviduct forms the white of the egg and the shell. These may be seen after the organs of the digestive system have been removed. The oviduct is the passage along which the eggs pass from the ovary to the cloaca. The yolk of the egg develops first, and this takes place in the ovary. The ovary will contain hundreds of very tiny yolks. These yolks will be found in all stages of development. Each yolk is enclosed in a sac or follicle, through which nourishment is obtained during development, such nutriment being supplied by the hen from the products of the food she has digested and being brought to the ovary by the blood. Each yolk is surrounded by a thin skin known as the vitelline membrane. This can be seen if the egg is broken into a cup. It prevents the yolk mixing with the white. On the surface of the yolk of a fertile egg a small circular area, the 'germinal spot', will be seen. This is the beginning of the development of the chick. If an egg is hard boiled and the white removed, a depression will be noticed in the inner surface of the white, and a

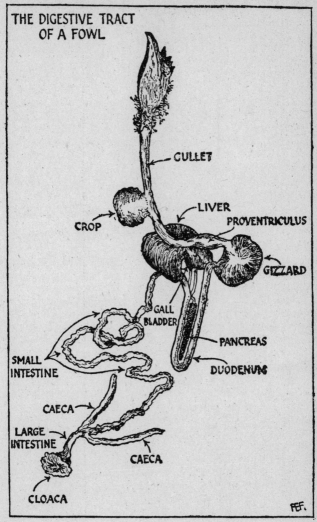

THE DIGESTIVE TRACT
OF A FOWL

GULLET

LIVER

CROP

PROVENTRICULUS

GIZZARD

GALL
BLADDER

PANCREAS

SMALL
INTESTINE

DUODENUM

CAECA

LARGE
INTESTINE

CAECA

CLOACA

FEF.

Fig. 11

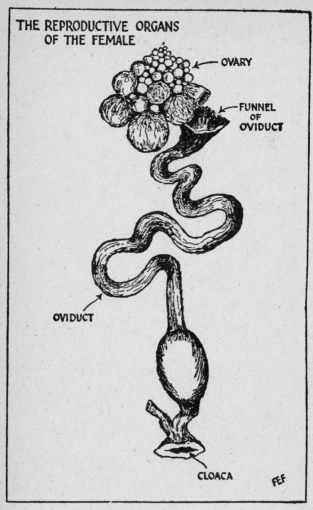

THE REPRODUCTIVE ORGANS
OF THE FEMALE

OVARY

FUNNEL
OF
OVIDUCT

OVIDUCT

CLOACA

FEF

Fig. 12

corresponding elevation appears on the yolk; this is the place where the actual chick is developing. When the yolk becomes ripe or mature, it breaks away from the covering of the ovary and enters the mouth of the oviduct. There it commences its passage through the oviduct, where various glands secrete the albumen—i.e. 'the white' of the egg. This is poured round the yolk, but owing to the rotation of the latter it becomes twisted at the ends, forming the chalazae. About 40 per cent of the albumen or white of the egg is supposed to be supplied as the yolk passes through the upper half of the oviduct.

After the yolk has passed some way down the oviduct, the shell membrane is produced by the membrane-secreting portion of the oviduct and a good deal more albumen is added. Here the egg is beginning to take on its final shape and size. It will have taken about three hours to pass through the first part of the oviduct and will remain from twelve to eighteen hours in the lower portion of the oviduct. From the membrane-secreting portion it passes into the shell gland, where the shell is laid on. Finally, the complete egg is expelled through the cloaca. The shell of the egg when first laid is moist, but it soon dries. As stated elsewhere, if two yolks enter the oviduct at the same time, a double-yolked egg is the result.

The egg

The hen egg is oval in shape, the usual weight being about two ounces. The shell will be white, brown or tinted. It is composed of carbonate of lime. If a fresh egg is broken into a dish or saucer, the chalazae and the layers of albumen can be seen. The chalazae are attached to opposite sides of the yolk and extend into the albumen towards the ends of the egg, or they may be twisted up close to the vitelline membrane. The chalazae, in the form of long threads, prevent any quick change in the position of the yolk and cause it to revolve on the long axis of the egg, keeping the germinal disc on the upper side nearest the heat in incubation. The dense layer of albumen is transparent and in a fresh egg can be seen, as well as the other two layers. The outer layer looks watery.

STRUCTURE OF AVIAN EGG

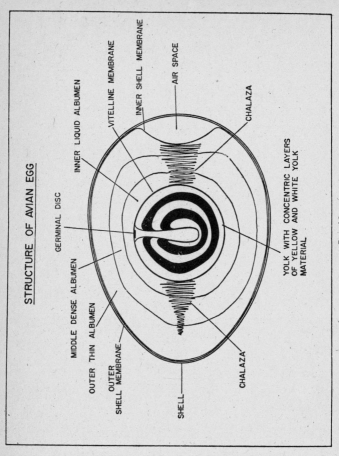

GERMINAL DISC

INNER LIQUID ALBUMEN

VITELLINE MEMBRANE

INNER SHELL MEMBRANE

AIR SPACE

MIDDLE DENSE ALBUMEN

OUTER THIN ALBUMEN

OUTER SHELL MEMBRANE

SHELL

CHALAZA

CHALAZA

YOLK WITH CONCENTRIC LAYERS OF YELLOW AND WHITE YOLK MATERIAL

Fig. 13

The shell membrane is double. In a perfectly new laid egg the shell is completely filled, but after it has been laid for some time and has got cool there is a certain amount of evaporation, and the double membrane separates into layers at the large end of the egg and the air-space is thus formed.

The composition of the hen's body, according to Cambridge data, is:

	Body including feathers	Plucked carcase
Protein (or flesh)	24·8%	19·3%
Fats	12·8%	13·5%
Ash	4·0%	4·4%
Water	58·4%	62·8%

The average composition of the new-laid egg, according to Cambridge data, is:

Yolk	33·1%
White	57·8%
Shell	9·1%

The chemical composition of the entire egg is:

Protein	10·9%
Fat	10·5%
Ash	0·9%
Water	68·6%
Shell	9·1%

Candling

All poultrymen should know enough about the make-up of an egg to tell a stale one from a fresh one, or an egg whose internal condition makes it undesirable to market or use it. One should also know the reasons why these undesirable qualities arise. The internal condition of an egg can be determined by what is termed candling. This is holding the egg in front of a bright light, preferably in a darkened room. A strong light is necessary, especially for brown-shelled eggs. The contents of the egg will be seen by placing it in front of the light and below the level of the eye. Before doing this, the egg should be given a quick twirl. This action will cause the albumen and yolk to rotate and the interior will be more easily observed. Holding the egg below the level of the eye enables the air space to be seen more readily. In fresh eggs the air space will only be small, about $\frac{1}{8}$ inch in depth. A large air space is an indication of staleness or age. A movable air space means a broken membrane, which will allow the yolk to stick to the shell.

The yolk of a fresh egg, if broken into a dish, will look well rounded, whereas the yolk of a stale egg will spread out and look flattened. When candling, the yolk should be only dimly visible as a shadow and should not move far from its normal position.

The white of an egg should be firm and clear. It is said that a watery or thin white is an indication of staleness. Eggs with blood and meat spots are sometimes noticed on candling. Blood spots are frequently found on the side of the yolk, but meat spots are generally present in the white. These egg faults are caused by the rupture of small blood vessels in the ovary.

Sometimes an egg will be found which on candling appears to have a greenish tinge. If it is broken open an olive-coloured yolk will be found. Very often this is caused where the fowls on range have discovered and eaten shepherd's purse or other weeds of a similar kind.

Shell-less eggs are produced as a result of not giving the hens a sufficient supply of oyster shell or limestone grit. Double-yolked eggs are not the result of disease but occur when two ova are passed from the ovary into the oviduct at approximately the same time.

Chapter 5

Incubation

Selection and handling of hatching eggs. Development of the chick (embryology). Physical requisites for successful incubation. Natural and artificial incubation. Incubators large and small. Hatching hygiene.

Hen eggs take twenty-one days to hatch out, duck eggs twenty-eight days, geese eggs thirty days, turkey eggs twenty-eight days, guinea fowl eggs and pheasant and partridge eggs twenty-four to twenty-five days.

Hatching by natural methods (broody hens) is not practical on commercial poultry farms today. Only small backyard poultry keepers and bantam breeders practice natural incubation. This is because the labour involved is high and only small numbers can be incubated.

Selection and handling of hatching eggs

To obtain best hatchability results several points in handling and selection should be adhered to. On average, about 15–22 per cent of all eggs incubated fail to hatch, thus any improvement will be an economic one.

The best hatchability results are obtained with hen eggs which weigh between 2 and 2¼ ounces, and since egg shape is inherited selection pressure on this characteristic will help to maintain good egg size.

Larger eggs than this may be a handicap in the setting trays, and because of their larger surface area probably have a higher physical requirement. The shape and shell texture of hatching eggs should be given careful consideration. Only those eggs with good smooth shells and a normal shape should be incubated. Long, pointed and round eggs are not successfully incubated. Naturally, cracked eggs must not be incubated.

Hatching eggs should be collected at least three times each day and held at a temperature of 60°F. Incubate them not later than ten days after collection.

Dirty hatching eggs should be dry cleaned with wire wool. If shell contamination is excessive, there may be no alternative but to wash them. In this case the water temperature should be between 80°–85°F and should contain a germicidal compound.

Handle hatching eggs with extreme care. Do not jar or shake them, for this may damage the internal contents and reduce hatchability.

Development of the chick

After fertilization has occurred in the hen's oviduct, the resultant zygote divides and develops by a process of segmentation. The single-celled zygote forms two, the two four, the four sixteen and so on, until a chick containing millions of cells has developed. Incubation has thus commenced before the egg is laid. The cells form three different layers: top, bottom and middle. Each layer of cells forms specific functions, such as skin, tissue and bone, the circulatory system and the nervous systems. It is these three layers which multiply to form the mature embryo.

Under the correct conditions of incubation, the legs and wings are discernible by the fourth day. Brain and nervous tissue have also commenced development. By the sixth day, the main subdivisions of the legs and wings are evident. By the ninth day, the embryo begins to look like a chick. Calcification of the bones is complete by the fifteenth day, and the colour of the chick's down is seen by the thirteenth day. The beak and claws are formed by the sixteenth day. On the nineteenth day, the yolk sac on which the chick grows is slowly drawn into the body cavity. The yolk sac will provide food material for the embryo for a few days after hatching.

Final development involves the muscle maturation in readiness for the strain of breaking or 'pipping' through the shell. When the shell is 'pipped' the embryo rotates within the egg until the top of the egg is chipped off. This enables the chick to escape from the shell prison where it has spent, in the case of a hen, the last twenty-one days.

Many embryos fail to hatch because they are in an incorrect position by the twenty-first day. This may be caused by careless handling of the eggs during the last week of incubation, insufficient turning and heredity.

The correct position for the embryo at the time of hatching is body along the axis of the egg and head towards the broad end, facing right and under the right wing. In this position the head will be facing the air space.

Some malpositions encountered are: head between the thighs; head in small end of the egg; head turned to the left; body across the egg; feet over head; and head over wing.

The most critical periods in the development of the embryo are between the third and fifth days, and between the eighteenth and nineteenth days. To a lesser extent the twelfth and fourteenth days are also critical.

Physical requisites for successful incubation

In artificial incubation there is one golden rule. This is 'Carry out the operation instructions advised by the incubator manufacturers'.

The following are the requirements for successful incubation: (a) correct temperature, (b) correct relative humidity and (c) correct ventilation.

Temperature

The normal hen temperature is between 105°F and 107°F. The optimum temperature for hatching determined at the centre of the egg is 100°F. The temperature at the surface of the egg is 103°F, and this is the temperature guide used in flat-type or still air incubators. In forced draught incubators, where the air is blown round the incubator by fans or a paddle, the temperature guide is 100°F. The difference of 3°F exists because, in natural draught incubators, the air is in layers and is at different temperatures as warm air rises. In forced draught incubators, there is no variation between the internal surface and external surface of the egg. Thus the temperature requirement is 100°F.

In large hatcheries the eggs are transferred to what are called separate hatching machines after eighteen days of incubation. The temperature of the separate hatcher is 99–99½°F. The slightly lower temperature is necessary because a certain amount of heat is generated by the hatching chick. In machines where the setting and hatching compartments are combined the temperature must remain at 100°F.

Relative humidity

The relative humidity is the amount of moisture in the atmosphere. This may be expressed as a percentage. A certain amount of moisture is required by the developing embryo because moisture is lost in the incubator. Too great a loss can cause the embryo to die. Too much moisture can cause the embryo to drown. Thus the requirements are critical to obtain good hatchability.

Relative humidity may be measured by comparison of dry bulb and wet bulb readings. At the normal temperatures and relative humidity recommended for large cabinet-type incubators the dry bulb (temperature) should be 100°F. and the wet bulb 88°F.

Ventilation

Ventilation is of great importance, for not only is adequate oxygen required but also harmful gases such as carbon dioxide must be quickly removed from the incubator. Ventilation therefore serves two purposes: to supply oxygen and to remove harmful gases. The incubator room must be well ventilated, for it is from here that the incubator draws its supply. Air inlets must be baffled to prevent draughts, and air should not be allowed to strike the sides of the incubator.

Position of the egg during incubation

The ideal position for the egg during incubation is either

lying flat on its side or with its broadest end facing upper-most. In both of these positions air is able to circulate freely and the embryo can orientate satisfactorily, thus avoiding malpositions. The former position is used with flat or still air machines and the latter with forced draught or cabinet-type machines.

When eggs are faced with their narrow or pointed end uppermost, many embryos adopt malpositions and die. This is because the beak fails to penetrate the air space at the broad end of the egg.

Turning hatching eggs

The more incubated eggs are turned through 180° during incubation, the better they usually hatch. Egg turning, which may be carried out by hand or automatically, prevents the embryo from adhering to the shell membranes. It also insures an even distribution of warm air.

The needs for turning are greatest during the first eighteen days of incubation. After eighteen days, turning should cease. When small flat-type incubators are used, hand turning is usually carried out three to five times each day. It must be done an odd number of times to prevent them spending the long night period on the same side for eighteen days.

With mechanical turning, usually effected every hour, this practice is not necessary.

Common reasons for failures in incubation

Trouble	*Probable reason*
(a) Delayed hatch.	(a) Setting stale eggs. Low incubator temperature.
(b) Early hatch.	(b) Temperature too high.
(c) Many embryos dead in shell.	(c) Incorrect turning of eggs. Faulty breeding stock. Disease. Nutrition at fault.
(d) Weakly chicks.	(d) Overheating of hatching unit.
(e) Sticky chicks.	(e) Temperature too high.
(f) Small chicks.	(f) Insufficient moisture. Setting too small an egg.

(g) Too many   (g) Insufficient male birds in breeding
    infertile eggs.   pen. Males too old. Hens too fat.
                            Eggs kept too long under ad-
                            verse storage conditions.

Natural incubation

Hatching under broody hens is ideal for the smallholder
raising only a dozen chicks or so each year. One broody hen
can usually incubate twelve to fifteen eggs. A broody hen is a
bird that remains on the nest box during the night and
ruffles her feathers at anyone's approach. She usually loses
feathers from under her wings and legs.

Before setting the broody hen, test her for two or three
days by letting her sit on artificial eggs (China eggs).

The ideal broody nest box measurement is 14 inches by 14
inches by 16 inches high. The centre of the nest should form
a saucer shape to prevent the eggs from rolling out. Hay or
straw makes good nest box material. Before setting the hen,
dust her with an insecticide to free her from external
parasites. Allow her to get off the nest once per day for about
ten minutes. This will give her time to feed, water and
empty her bowels. She should not be allowed to leave her
nest for longer than fifteen minutes each day. As soon as the
hatch is off, all the debris, such as egg shells, should be
removed. Unhatched eggs should also be removed and disposed of. Replace dirty litter material. The broody hen
should be dusted again with an insecticide so that she can
comfortably brood her chicks.

Incubators

Flat-type or natural draught incubators.

A clear idea of the proper working of an incubator is an
absolute necessity for anyone embarking on this work.

The accompanying illustration shows the cross-section of
a small unit hot air machine.

This type of incubator is heated by hot air. They either
stand on their own legs or on a flat surface. Heat is produced
and circulated by a lamp (L); the fumes rise up a central flue

(F) and down the outside of the fresh air chamber, and are diffused in the incubator room. Fresh air only gains entry to the machine's interior. Heated air passes into the machine, where it diffuses through a canvas diaphragm above the egg tray (E). The air passes over the eggs and out through the ports at the bottom of the machine (P).

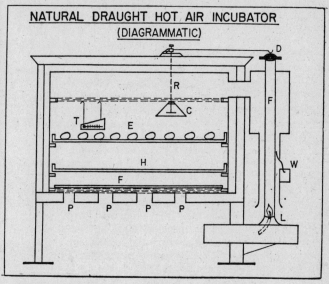

NATURAL DRAUGHT HOT AIR INCUBATOR
(DIAGRAMMATIC)

Fig. 14

Humidity is provided by a wick (W) dipping into water. Being suspended near the cool fresh air flow it provides moisture to the incubator atmosphere.

Ventilation is provided by felts (F) situated at the bottom of the machine. There are two felts. Two are normally present at the commencement of incubation.

The first felt is removed by the tenth day and the second by the nineteenth day of incubation. The thermometer is suspended 2 inches above the egg tray and the bottom should just touch the eggs. Temperature control is governed by a small capsule (C) containing ether, which as it expands pushes the rod (R) upwards and thus raises the damper (D). Excess heat is therefore allowed to escape. A reduction in temperature causes the capsule to contract and lower the damper bar and rod.

A hatching tray (H) is situated below the egg tray, into which the chicks fall as they are hatched.

This type of machine may hold as few as fifty eggs and as many as 250, depending on cabinet size.

Cabinet machine (forced draught)

Many more eggs may be incubated in cabinet machines in a much smaller amount of space. Modern machines automatically turn their eggs every hour.

Most large machines are constructed of a stout insulated casing which contains layers of non-conductive materials, e.g. fibre glass. The internal surface is usually zinc for ease of cleaning and also because it is non-corrosive.

Moisture may be provided by large water pans situated at the bottom of the machine. In large so-called 'walk-in-type' incubators, in which the attendant is able to walk, stand and work, moisture is sprayed into the incubator by water jets.

Heat is provided by electricity, although paraffin models can often be purchased. Electric bars may be positioned centrally but are always close to the heat circulator, which may be a fan or paddle. The fan or paddle propels the air evenly around the cabinet, passing it over and under the incubating eggs. Stale air is forced out either through the top or through the sides of the incubator.

Temperature is controlled by a thermostat, which cuts in and out depending upon temperature. In case of emergency an alarm system is generally incorporated into the electric circuit. Figure 15 shows a cabinet incubator which is used for setting eggs to the eighteenth day only. On the eighteenth day the eggs are 'candled', and all infertile eggs and dead germs are removed. They are then transferred to a hatching incubator for the remaining three days of incubation. Using what are known as separate setting and hatching machines allows each to be carefully disinfected and cleaned after each hatch, and at the same time increases the hatchability.

FORCED DRAUGHT CABINET INCUBATOR
(DIAGRAMMATIC)

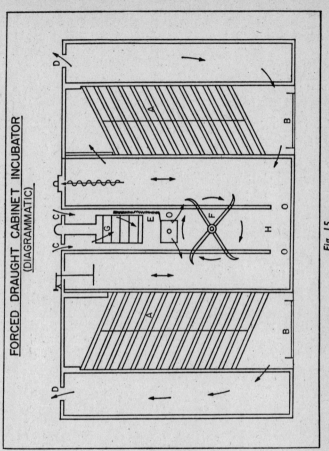

Fig. 15

A = Setting trays
B = Water trays
C = Air inlet
D = Air outlet
E = Heater
F = Fan
G = Hot water
    radiator
H = Central heating
    in mixing
    chamber

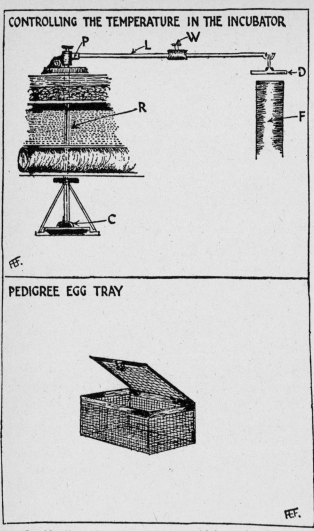

CONTROLLING THE TEMPERATURE IN THE INCUBATOR

PEDIGREE EGG TRAY

Fig. 16 Above: A most important part of the machine.
Below: Used for keeping track of special eggs during incubation.

The percentage hatchability is the number of chicks which hatch, divided by the original number incubated and multiplied by 100. Under good conditions this figure should be around the 80 per cent mark.

Hatching hygiene

The hatchery should be kept as clean as a maternity hospital. It merits the most meticulous attention to hygiene at every stage of its operations.

Infection which causes diseases may enter the hatchery by four main routes. These are:

(a) Hatching operators.
(b) Fleas.
(c) Vermin and other external parasites.
(d) On the surface of dirty eggs.

Eggs should be produced from breeding birds known to be clear of egg transmissible diseases. The operators must wash ther hands periodically through the day. Sexers visiting the hatching must disinfect themselves and their clothing both on arrival and on leaving. The use of machines which spray aerosol sprays are thoroughly recommended. They help to keep down the 'bacteria count' in the building, although they do not replace disinfection.

All incubator equipment must be treated with boiling water containing a 4 per cent solution of washing soda. Fumigation is best done with the chemicals potassium permanganate and formalin. Rubber gloves should be worn when disinfecting with these chemicals because they react violently when mixed.

For each 100 cubic feet of incubator space the following amounts of these two chemicals should be used:

Formalin, $4\frac{1}{2}$ fluid ounces.
Potassium permanganate, 3 ounces.

The incubator room itself should be kept spotless, and all debris such as egg shells and dead chicks should be burnt in an incinerator.

Transferring the chicks to the brooders

Where the poultry farmer hatches and broods his own chicks, this operation should be carefully carried out. Normal chick boxes should be used, taking care not to over-fill the boxes. Chilling the chicks must be avoided at all costs.

Maintaining incubator records is vitally important, for analysis of them may often enable the hatcheryman to be more efficient. Eggs incubated, fertility, hatchability and cull chicks should all be recorded and entered on a proper filing system. Scraps of paper are useless for this purpose. The keeping of accurate records will help to put a finger on causes of trouble.

Chapter 6

The Growing Chick

Brooding—natural and artificial methods. Different makes of brooders and their management. Management of growing birds. The identification of sexes and culling.

Rearing is one of the most important parts, if not the most important part, of the poultry farmer's work. The rearing season, which was traditionally in the spring months of the year, is now practised all the year round. This has been due to the introduction of intensive rearing methods and houses in which the environment is almost completely controlled. Rearing has a great influence on the future performance of the laying stock. The rearing period runs from the day the chick is hatched right through the time it goes into the laying quarters or, in the case of the table bird, up to the time it is killed.

The first part of the rearing is known as the brooding period. This is the time in which the young chick must be provided with some warmth, either naturally from the mother hen or artificially from brooders.

Coal burning stoves and hot water pipe brooders are not used much today, and wherever possible electricity should and has taken the place of all types because of the great saving in labour and the reduction of fire risks. The infrared brooders are very efficient. Calor gas is very satisfactory when electricity is not available. Brooders of both gas and electricity have chick capacities up to 1000 birds or so.

The period when heat is required alters slightly according to the time of year. Six weeks is the usual time, although during the summer five weeks will be adequate.

The second part of rearing is after the chicks leave the

brooders and are transferred to grow on in range shelters, carry-on cages or intensive rearing houses.

The brooding period

Baby chicks, unlike most young animals, are unable to live for long in the normal temperature. After a short time outside they must be able to go somewhere to warm up. If they become chilled they will either die or grow stunted. A slightly chilled chick often develops digestive disorders.

Rearing depends on the combinations of three factors. These are: (a) good management, (b) quality of stock and (c) correct nutrition.

It is false economy to buy and rear the cheapest stock available. Chicks which appear to be sick or ailing should not be accepted. Chicks should be the progeny of blood-tested stock. This precaution is a safeguard against the disease Bacillary White Diarrhoea (B.W.D.).

Good management is essential for success to be achieved. Much commonsense is required.

Good nutrition is essential during the birds' first few weeks of life. Food consumption is small during the first six to eight weeks, so only good-quality, palatable food should be given.

Natural brooding

Hens are not used on a large scale today for rearing chicks. Only with very small poultry units are they considered. This is because large numbers of broody hens are difficult to obtain. The labour entailed in natural brooding is also very high, as each bird generally only rears twelve to fifteen chicks.

The rearing coop should measure 2 feet by 2 feet by 2 feet high. It must be weather- and vermin-proof. The hen is confined to the coop by means of vertical bars 2 inches apart. These permit the chicks to escape from the hen and yet allow the bird to put its head through to feed. Litter such as hay, wood shavings or peat moss should be used on the floor of the rearing coop.

Food and water must be constantly available and a good-quality chick mash is ideal. The water container must only be deep enough to allow the chicks to drink. Open water troughs are not recommended for fear of drowning the chicks. The whole coop should be surrounded by wire netting to prevent the chicks wandering too far from the mother.

CHICKEN COOP WITH RUN

FEF.

Fig. 17 For mother hen with her chicks.

The age at which the hen can be taken away from her brood is a matter for observation. She will be beginning to tire of them when they are about six to ten weeks old. She can then be returned to the laying flock and the chicks transferred to a rearing house for *larger carry-on brooder*.

Artificial brooding

Artificial brooding refers to the use of equipment which provides conditions similar to those of the broody hen.

There are many methods of artificial brooding, varying from small paraffin heat pyramid brooders which rear fifty to seventy-five chicks to large calor gas and electric brooders which rear several thousand, such as those used in modern broiler production.

C

Temperature requirements

The correct brooding temperature is between 95°F and 100°F measured 2 inches above the litter level. As the chicks grow, so the temperature requirement decreases. The following figures should be used as a guide to brooding temperatures.

Age in weeks	°F
Day-old to 1 week	95
1–2	90
2–3	85
3–4	80
4–5	75
5–6	70
6–7	65

The best guide to temperature is the state or condition of the chicks. If they are huddled around the heat source, then they are too cold. On the other hand, if they are spread right away from the heat, then they are too hot. Huddling in one corner suggests a draught on the opposite side. Contented chicks will be spread evenly over the heated area. Manufacturer's instruction supplied with every brooder should be followed.

Source of heat

Heat for brooding may be supplied by electricity, gas, solid fuel, paraffin oil or hot water.

Electricity is the most labour-saving and simplest to operate, although in some areas the source may be unreliable. Under these circumstances gas makes a very good substitute.

There are many types of electric brooders on the market, varying from small infra-red units to large heat storage units.

Solid fuel brooders are little used today because of the high initial capital outlay of providing a boiler and heat transfer pipes around the brooder house.

Likewise, paraffin is little used, although some poultry

keepers keep it as a standby in unreliable electricity areas.

Infra-red brooders

These are two types: dull emitter and bright emitter. As their names suggest one form gives a dull heat, the other a bright heat. Both are equally as good, although the bright

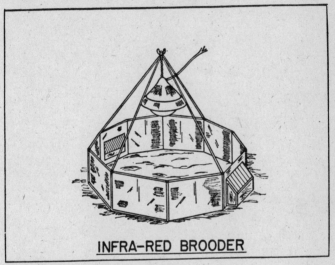

INFRA-RED BROODER

Fig. 18

emitter does allow the attendant to observe his chicks without disturbing them. Infra-red lamps usually have 250-watt bulbs, which are suspended some 18 inches to 24 inches from floor level. As the chicks grow, so the height may gradually be increased to 30 inches. The only other equipment required is metal or wooden surrounds, which confine the chicks to the heat source. These are gradually moved away as the birds grow. Each 250-watt lamp is sufficient for brooding seventy-five chicks up to seven weeks of age.

Fig. 19 An Infra-Red and large Calor Gas Brooder.

Tier battery brooders

This system consists of tiers of cages placed one upon another, with a space between each to allow for manure collection. They are usually five tiers high and may be operated by electricity or calor gas. The chicks are brooded in the cages for three to four weeks, after which they may be transferred to hay box brooders or cooler cages. A typical tier brooder is shown in Figure 20. About 1 square foot

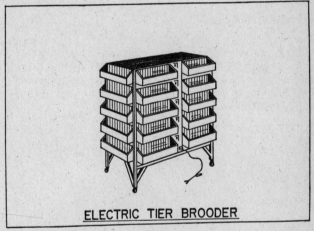

ELECTRIC TIER BROODER

Fig. 20

of brooder space is allocated to each six chicks. The heating pad or unit may be placed centrally or at one end. There are usually eighty chicks to each tier; thus a five-tier unit will brood 500 birds to three or four weeks of age. This brooding system saves a considerable amount of space compared with floor rearing methods, although the complete unit is more expensive than its equivalent floor rearing equipment. A brooder house is required to house the tier brooder. This should provide a room temperature of about 65°F–70°F.

Warm floor brooders

This type of brooder consists of a unit which measures 3 feet by 2 feet. It has boarded sides and ends. Food and water troughs are fixed to the sides, and a pophole at one end allows the chicks to escape from the heated area. The floor consists of ½-inch mesh wire, above which a blanket type canopy is fitted. The heater units are under the wire floor. Small paraffin 'putnam' stoves are normally used for heating.

Free range brooding

Despite the increasing popularity of the intensive rearing systems, range rearing is still carried out, particularly on small poultry farms. There is no doubt that during the spring and summer months chicks benefit from being reared on good pasture. Feathering is generally very good. Intensive rearing is generally practised for the first four weeks, after

FIELD HOUSE

Fig. 21   A good house for the stubble.

which the chicks are transferred to hay box or fireless brooders. The hay box measures 3 feet by 9 feet long. One end, 3 feet by 3 feet is the brooding or cooling-off compartment and consists of a rack 10 inches from floor level on which hay bags are placed to retain liberated body heat. Each unit holds thirty-five growing birds to eight weeks of age. Feed and water troughs are generally placed in the 3 feet by 6 feet run, although some hay boxes have feed troughs in the sleeping compartment.

This type of brooder requires daily moving, thus level ground is essential.

SUSSEX NIGHT ARK

Fig. 22

After eight weeks the birds may be transferred to night arks or range shelters. A night ark is 6 feet long by 3 feet wide. Each holds twenty-five growing birds from eight weeks to point of lay. They require moving daily to prevent the grass becoming 'poached'.

Range shelters are similar to night arks but have a roof which reaches nearly to the ground. The ends and sides are covered in wire mesh 2 inches in diameter, whilst the night ark has wooden boarded sides. Each unit houses fifty to eighty growing birds from eight weeks to point of lay. Range shelters may also be used for rearing stock cockerels to maturity.

Management of growing birds

The rearing house must be dry and draught-free. It should also be well ventilated. Before the chicks arrive, light the brooder to achieve the required temperature.

Sufficient food and water trough space must be provided so that each chick can feed and drink in comfort.

RECOMMENDED FOOD SPACE PER 100 CHICKS

Age in weeks	Mash feeding (ad lib)
0–2	8ft
2–8	12 ft
over 8	20–30 ft

FLOOR SPACE PER 100 CHICKS (INTENSIVELY REARED)

Age in weeks	Area of floor
0–4	40 sq. ft
4–8	100 sq. ft
8–24	120–250 sq. ft

After eight weeks of age the chicks should be changed from a chick food onto a growing diet. If they are being intensively reared, then an intensive grower's ration should be used. Feeding should be *ad lib* throughout, and small quantities of grain may be fed. Sometimes a grain balancer-type growing ration is used, especially if the farmer has surplus cereals. It is essential to follow the feedingstuff manufacturer's instructions, for overfeeding grain may cause nutritional deficiencies. Insoluble grit, such as granite and flint, should be fed in token amounts. It will help the growing bird to digest its food. It is important to feed the correct size, because if it is too small it can cause digestive disturbances.

The sexes

Growing stock must on no account be overcrowded, for this will lead to bullying and result in cull birds. The sexes should be separated by eight to nine weeks of age. The males have larger combs and wattles than the females, and also better-developed tail feathers. The females have a much smaller bone structure than the cockerels. Separating the

sexes will prevent bullying and if necessary the cockerels
can be fattened for market. Any 'cull' birds or passengers
should be removed from the flock. To leave them only
increases the disease risk. They also consume valuable feed
and occupy space which can profitably be used by other
growing birds. As the pullets approach maturity, they should
be handled extremely carefully. When catching them, do so
by putting the fingers of one hand between the legs. This
will prevent the developing internal organs from being
damaged.

Culling

Culling should be a continuous practice on the poultry
farm. It should be commenced directly the chicks have been
hatched. Any chicks with unhealed navels, deformities and
lack of vigour should be killed. Only healthy chicks should
be placed under the brooder. Diseased growing stock must
not be tolerated. Their presence in an otherwise healthy
flock should not be tolerated for they make the flock more
susceptible to disease. Stunted birds should be removed and
marketed, or if too small they should be destroyed.

Each time the growing stock is handled or moved into
alternative rearing accommodation it should be checked for
signs of disease. This is particularly important when
housing-up point of lay pullets. Badly feathered birds may
reflect on the feeding and environment.

Intensive rearing

Many of the large-scale pullet rearers house pullets under
extreme intensive conditions. The pullets are reared in
houses which are similar to those used in broiler production
Some pullet rearers utilize the deep litter system, whilst
others use wire and slatted wooden floors for rearing. In
these two latter housing conditions, the birds do not come
in contact with their droppings and thus a higher stocking
rate is permissible. Disease risks are also reduced. In the so-
called controlled environment rearing houses, both the
lighting and the ventilation are completely artificially controlled.

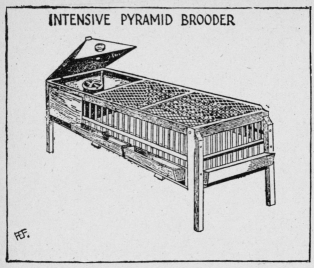

INTENSIVE PYRAMID BROODER

Fig. 23 This type of brooder gives excellent results.

Food consumption guide (cumulative) per 100 pullets

Age (weeks)	Pounds food consumed	Age (weeks)	Pounds food consumed
1	22	15	1300
2	45	16	1450
3	80	17	1500
4	125	18	1700
5	200	19	1800
6	275	20	2000
7	380	21	2100
8	460	22	2200
9	550	23	2300
10	660	24	2450
11	760	25	2600
12	860	26	2750
13	1000	28	2900
14	1120	30	3200

Lighting growing pullets

Pullets which are reared in the winter and spring months mature earlier than those reared after midsummer. This is because a long day-length, increasing up to seventeen hours (midsummer), encourages early sexual maturity and short day-lengths, decreasing down to eight hours, delay sexual maturity.

Knowledge and use of these facts enables the pullet rearer to control the development of his birds so that he can accurately predict the age at which his birds will commence laying.

Under housing conditions, called controlled environment, where both light and ventilation is artificially maintained, it is possible to rear growing pullets under periods of very short day-length. For example, six or eight hours light in every twenty-four hours. It has been found that these short day-lengths in rearing can result in improved laying performance, because at point of lay the light is gradually increased by thirty minutes each week, thus simulating the spring light pattern. Where possible, an increasing day-length should be avoided with growing pullets as it tends to encourage early sexual maturity, very small eggs and the risk of prolapse occurring.

Under free range conditions in spring time early sexual maturity is difficult to prevent, but it can be modified to a certain extent by giving the birds supplementary light from a few weeks of age. If, for example, the pullets are spring hatched and would mature in August when the natural day-length is still long, their day could be extended to twenty hours by artificial means, and by small weekly decreases it could be lowered to the level of daylight expected in August. This light pattern would help control maturity and improve laying performance.

Colour of light

The lower the intensity of light, regardless of colour, the less will be the problem from cannibalism and feather pecking. Red light is no better in this respect than normal white light. Blue light must never be used as it tends to immobilize the birds. For rearing pullets intensively in windowless houses a light intensity of 1 foot candle is adequate.

Chapter 7

The Laying Bird

Precocity, lighting the laying pullet. Preparation of the laying quarters. Debeaking. Battery cages. Second season layers. Culling.

Precocity

When both light and heavy breeds are hatched at the same time, it will be found that the light breeds or types will mature about one month before the heavy breeds or types. Some light breeds will commence laying at four months of age, whilst some heavy breeds will not lay until seven months old. Early sexual maturity is not always a good thing because frequently many small eggs are laid initially. At the same time, late sexual maturity is not to be encouraged as this shortens the productive period. The poultry farmer should aim to bring his pullets into lay at about twenty to twenty-two weeks of age. As has already been mentioned, increasing hours of daylight encourages early sexual maturity; thus birds hatched during the winter and spring, if reared under natural daylight conditions, will mature early. To avoid this controlled environment rearing is practised, which can be used to control the age at sexual maturity. In this way small eggs may be prevented.

Early sexual maturity should not be delayed by the use of feedingstuffs unless in the hands of a very experienced and competent poultryman. Even then the system has its drawbacks.

With stock such as broiler-breeder females the combination of a controlled lighting programme and restricted feeding is practised. Such programmes are essential because the number of eggs laid by this class of stock is small and there is a requirement for a minimum size of hatching egg.

Lighting the laying bird

After the age of eighteen weeks, pullets must on no account have the amount of natural or artificial light reduced. This is important for, should it occur, then the onset of egg production will be delayed with possibly disastrous results to the laying potential of the bird.

Once egg production has commenced, the amount of lighting should be increased by artificial means each week until a level of seventeen hours is reached. The normal weekly increment is between fifteen and thirty minutes. To allow the light to decrease by only one hour can cause a check in production, with the likelihood of a neck moult occurring. Should a neck moult occur, then production is likely to be permanently damaged.

Artificial light may be provided in many ways. Electricity, gas and paraffin may all be used. Electricity is the most common because of its saving in labour and ease of use.

The light can be profitably supplied in the morning or evening, or a combination of both. Morning light is best supplied to birds on deep litter or other floor-laying systems involving deep litter or houses with perches. With morning lighting there is no fear of the birds being left on the floor when the lights are turned out. This could happen with evening lights unless a dimming device is used which allows the birds to retire to the perches or the droppings pit before the lights are finally extinguished.

Under the battery cage system this device is not necessary as the birds cannot, of course, leave their cages.

In areas where electricity is unreliable, paraffin lamps may be used. The lamps should be filled with sufficient oil to last a predetermined time. In this way the lamps only have to be lighted and not extinguished.

Artificial light not only prevents neck moults but also increases annual egg yield and, since the best egg prices are obtained during the late summer, autumn and winter months, profits are also higher when compared with non-lighted laying stock.

Feeding during the extra light hours is required to allow for the additional eggs laid.

Preparation of laying quarters

Before point of lay pullets are housed, their future laying quarters should have been completely prepared. The disinfection procedure, which should have been completed before the new birds were housed, should make provision for new litter material in deep litter houses, preservation of all timber and servicing of all automatic equipment. The birds should be housed some three weeks before eggs are expected in order to give them ample opportunity to settle down in their new surroundings.

Where perching facilities are made available in the laying quarters, it is necessary that the birds should be taught how to use them. This is particularly important if no such facilities were available in the rearing stage. For the first few nights it will be found necessary to place the pullets on the perches, removing any birds sleeping in nest boxes. The provision of perches or a droppings pit ensures that all the droppings fall in a certain area. This helps to keep the remainder of the litter in a dry condition and is therefore more comfortable to both birds and attendant.

The best size for perches is 2 inches by 2 inches, with the uppermost edges bevelled. They should be 18 inches away from any wall, and there should be 15 inches between each row of perches. The best height from floor level is about 30 inches. Each bird should be allowed 8 inches of perch space.

Nest boxes

Allow 1 square foot of nest box space to every four to five pullets. Each nest should measure 12 inches square and 14 inches high. They may be made in tiers, in which case an alighting perch should be attached to the front of each nest. Communal nests are generally 10 feet long by 2 feet wide. They have a bird capacity of eighty to one hundred. At each end and in the middle are small popholes to allow the birds to enter and leave. All nests should be at least 9 inches above litter level and should preferably have a wire floor to allow for good air circulation. Nest litter material, e.g. straw, wood shavings, sawdust and

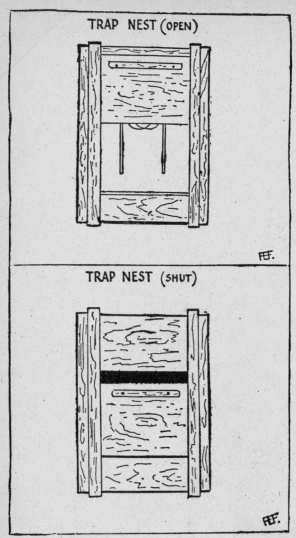

TRAP NEST (OPEN)

TRAP NEST (SHUT)

Fig. 24   Necessary if egg records are kept.

peat moss, should be changed when soiled. Eggs should be collected at least three times each day to prevent them becoming soiled or broken.

Broody coops

A broody coop may be any size. It is constructed of 2 inches by $\frac{1}{2}$-inch wooden battens spaced 2 inches apart. Food and water troughs are fixed to the coop front. Allow broody space on the basis of 5 per cent of the flock size.

Food and water hoppers

These should be spaced evenly throughout the house. Allow 40 feet of food trough space and 12 feet of water space per 100 pullets. In calculating the amount of space provided by circular hoppers, multiply the pan diameter by three. Food hoppers should not be filled to more than two-thirds their capacity and should be placed so that the top of the trough is level with that of the bird's back. In this way food wastage will be reduced.

The most popular way of feeding intensively housed laying stock is *ad lib*. This means that the food is constantly in front of them so they may feed at any time of the day. With the *ad lib*. system dry mash is best, although in controlled environment poultry houses pellets and crumbs may be used. Crumbs are pellets broken into small pieces.

Wet mash feeding may be used with free range birds but is not recommended for intensively housed layers because of the risks of cannibalism and the high labour costs.

Debeaking

Debeaking is the term used for the removal of part of the upper mandible (beak). Normally one-quarter is removed and the lower beak lightly cut back, say one-eighth of an inch. Debeaking is carried out to prevent vices occurring. The effect of having the lower beak longer than the top prevents the bird from injuring its neighbour. Usually electric debeaking instruments are used which cauterize the blood

DEBEAKING

NON-DEBEAKED   CORRECTLY DEBEAKED   BADLY DEBEAKED

Fig. 25

vessels. This prevents bleeding. Debeaking should be carried out carefully to prevent undue disturbance to the birds. The best time to do it is when the pullets are being housed up in their laying quarters. It should not be done as a matter of routine but only where there are risks of cannibalism. Handle the birds quietly and with gentleness. The diagram shows the correct amount of beak to remove.

Regular inspection of laying stock

Feather pecking and cannibalism frequently occur through some error in management. For example, the poultryman may overlook an injured bird or a pullet may catch its cob in wire netting, causing it to bleed. Unless action is taken cannibalism may start.

Birds badly infested with mite and body lice may peck themselves in order to be rid of the irritation. Frequent inspection for these parasites should be made and, if necessary, some action taken. The trained poultryman will handle his birds periodically to examine them for body condition and for the presence of external parasites.

Laying birds in battery cages

Pullets housed in battery cages usually settle down fairly quickly when first placed in laying cages, provided they are moved prior to the onset of production.

The removal of the less satisfactory birds should be a regular routine job in the battery house, and it is carried out mainly by observing which cages have laid the least number of eggs. Exact rules for culling cannot be prepared. Generally it is best to allow the birds at least two months to settle down before culling or attempting to cull the poor egg production. Of course, unhealthy birds must be removed immediately.

Regular inspection of the records will indicate to the poultryman which cages are performing best and which are not doing so well.

Eggs laid in battery cages do not need such frequent collection as those laid in the deep litter house. Twice a day is generally sufficient. Dusting the egg-collecting cradle will prevent the eggs from becoming marked by the cage wires. Water troughs should be cleaned out once each week to prevent stale food accumulating in the bottom of the troughs. Where pellets are fed to battery birds, the need for this cleaning may be less frequent.

The wisdom of maintaining a laying flock for a second year production

Unless the poultry keeper is in a very small way it is generally uneconomical to carry laying birds on for a second season's laying. This is because the egg yield is generally lower, the shell quality much poorer, the appetite larger and the second season's laying season shorter. Against these objections, one may say that as the flock is kept for two years no new pullets need be purchased. This is, of course, very true, although it should be remembered that a bird needs a pause in production between two laying seasons. This pause may be for a period of eight to ten weeks while the bird moults and regains its general body condition. This pause costs money and, whilst the cost is less than that involved in buying or rearing new pullets, the other disadvantages mentioned above should be borne in mind. The exact time of the year and state of the egg market must also be taken into account. However, if it is decided to maintain the flock for a second season's production, the following procedure is recommended.

Force moulting

The birds should be forced into a moult. This will cause an immediate drop in production but is essential if all the birds are to commence laying again at the same time.

To force moult withdraw both food and water for twenty-four hours, and completely black out the house. After twenty-four hours, feed 6 pounds of wheat per 100 birds for ten to fourteen days and increase the daylight to six hours. By this time the feathers should be dropping. Increase the feed to 9 pounds after three weeks and then feed the moulted birds some layers' food by the end of the fourth week. Increase day-length to nine hours. The laying food may then be fed *ad lib*. When eggs are required, return the birds to their clean and disinfected laying quarters, and increase the amount of light to fourteen or fifteen hours. Some birds refuse to stop laying even with this treatment. They should be forced again.

Force moulting is sometimes used for breeding stock which may be retained for two or even three years. The same procedure outlined above should be followed.

Culling

The selection and rejection of birds on the poultry farm must be a continuous process. Culling, which means the removal of unprofitable birds from the flock, is not a simple matter. Fortunately, today with modern high egg producing strains available little culling is required, as flocks are now much more genetically uniform. That it pays to cull some birds is obvious, for it must be remembered that the elimination of a poor layer means a reduction in the food bill.

For laying poultry today to be a reasonable paying proposition each bird must lay about 230–240 eggs in fifty-two weeks. The higher the figure, the better the birds will pay.

Culling really only eliminates those birds not laying at the time of handling. It tells the poultryman nothing about past or future perfomance. It must, therefore, be a hazardous business.

To determine whether a bird is in fact laying it must be handled and examined. The width of the two pelvic bones near the vent will tell the poultryman whether or not an egg has room to be laid. The bird which has a moist vent and a large space, say two fingers, between the pelvic bones will almost certainly be laying. The headgear, that is the comb, wattles and general head appearance, should be red and waxy. If dull, shrivelled and pale, the bird is probably not laying. The presence of moulting may indicate that a bird is just going out of lay or is, in fact, out of lay.

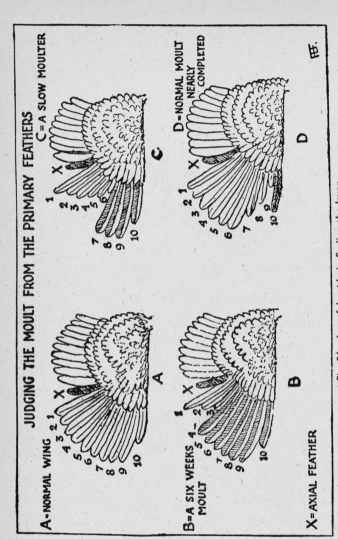

Fig. 26 A useful guide in finding the layer.

Birds producing most eggs are generally late moulters, i.e. they have laid for much of the laying period. Early moulters may be regarded as poor laying birds. Moulting commences at the head and neck, and proceeds down the body. The last feathers to moult are those of the wings. Thus it is possible to know if a bird has just commenced moulting or is almost through it (see Figure 26).

The plumage

The condition of the plumage gives some measure of a guide to the past performance of a laying bird. The feathering of a good layer will usually have a dilapidated appearance, caused by many visits to the nest box. Early moulters will, on the other hand, show signs of new and fresh-looking feathers. New quill feathers are large and sappy in appearance, the old ones hard and hollow. It is possible to estimate when the moult began by counting the primary feathers in the wing. Primary feathers are those at the outer end of the wing, whilst those known as the secondaries are on the inner part of the wing. The primaries and secondaries are divided by an axial feather which is much shorter. There are ten primaries and fourteen secondaries. The primaries are always moulted in numerical order. The high producing bird moults rapidly, the poor producer moults slowly.

In estimating when the moult began, allow six weeks for the first complete new primary. A wing having three new full-grown primaries will show the bird has been moulting for about ten weeks. If none of the primaries is fully grown, the estimate must be made allowing for this, on the basis that two-thirds of the growth is made during the first three weeks and the other third during the remaining three weeks. So a half-grown primary would show about two weeks' growth (see Figure 26).

Handling in preparation for culling

Poultry are usually nervous creatures and rough treatment may result in a drop in egg yield. In order not to upset the birds they should be handled carefully. To do this

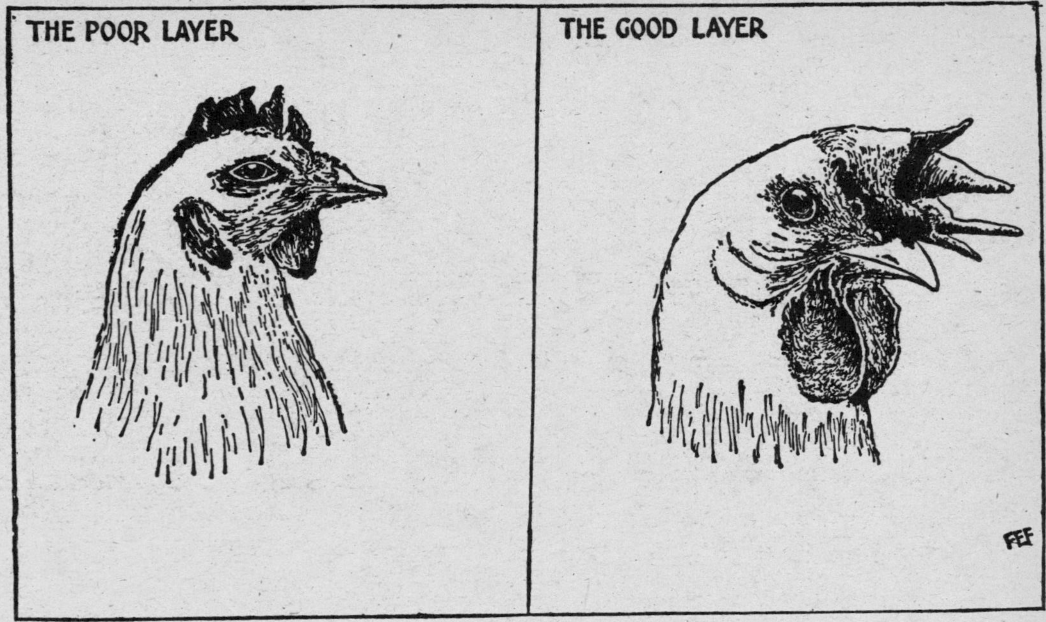

THE POOR LAYER | THE GOOD LAYER

Shrivelled comb and dull expression.   Bright eye and well-developed comb

Fig. 27

culling may be done at night, although an efficient job can be made of culling during daylight hours. With extensively and semi-intensively housed poultry, the use of a catching crate will prove very useful. The dimensions of the catching crates are 5 feet long by 2 feet wide by 2 feet high. At one end is a trap door and a sliding partition. About twenty birds may be driven from the house into the crate and the top end closed. The birds may be handled by removing them via the trap door built into the crate top.

Deep litter house culling may be carried out by dividing the house into two sections with wire netting after all the birds have been driven into one side. A smaller wired area may be used to confine and inspect a few birds at a time.

Rejected birds may be transferred to an empty poultry crate, whilst those passing the inspection should be placed on the other side of the divided house.

Random sample laying trials

The breeders of hybrid laying birds enter their stock in laying trials in order to assess performance at an independent and unbiased centre. At these centres all aspects of production, such as egg numbers, size, shell quality, food consumption and liveability, are carefully recorded. Egg producers wishing to assess comparative performance between the many strains are able to examine carefully the records which are published periodically. The producer is advised to use the information, together with results produced by birds of the same breed in his locality, for decision making on which breed to use relative to his requirements.

Chapter 8

Feeding and Nutrition

*Introduction. Proteins, carbohydrates, fats, water,
minerals and vitamins. Antibiotics, energy protein
ratios, food requirements of the bird and specimen
rations, feed ingredients, practical feeding.*

Correct feeding is extremely important in good poultry
keeping.

Food costs amount to approximately two-thirds of the
total costs in producing poultry meat and eggs. Therefore,
however good the stock, however efficient the management,
injudicious, extravagant or wasteful feeding practices can
easily be the means of reversing a potential profit into a loss.

The poultry farmer must have a good working knowledge
of feeding and nutrition. He should be able to recognize the
common ingredients at a glance and be able to differentiate
between a good and bad sample.

There is much to be said for purchasing proprietary
feedingstuffs of excellent quality from firms of reputation.
Much time and worry is saved by purchasing ready com-
pounded rations. Home mixing requires a great deal of skill
and labour, and should only be attempted by large poultry
keepers who have the ability and knowledge: knowledge
because different ages of stock and classes have very different
nutritional requirements, and ability because unskilled
labour may omit essential ingredients, to the detriment of the
stock and the poultryman's livelihood. Nutrients are required
for many purposes: (a) to maintain a healthy body condition,
(b) to produce maximum growth rate, (c) to provide for
the production of eggs and (d) to allow a satisfactory
growth rate (in the case of growing pullets).

The first requirement is known as the maintenance requirement. If only this is provided very few eggs would be

produced and growth rate would be sub-optimum. To produce eggs and flesh a production ration is required. Likewise, all young growing stock require a certain daily amount of food for growth, in addition to that necessary for maintenance. Growth can only take place when the ration provides nutrients in excess of the bird's maintenance requirements.

Foodstuffs may be classed as nitrogenous suppliers and non-nitrogenous suppliers. Nitrogenous suppliers contain protein in the form of amino-acids, whilst non-nitrogenous suppliers provide oils, fats and carbohydrates.

Foodstuffs are made up of: (1) proteins, (2) carbohydrates (energy), (3) fats and oils, (4) water, (5) minerals and (6) vitamins.

Proteins

Proteins or flesh formers are found in abundance in animal products such as fish meal, meat and bone meal, and milk. They are also abundant to a slightly lesser degree in vegetable products such as soya bean meal, groundnut meal, peas and beans. To a much lesser degree they are found in cereal grains and their by-products. They are essential food nutrients, both for production and for maintenance. Proteins contain the elements carbon, hydrogen, oxygen, nitrogen, phosphorus and sulphur.

Proteins are built up from amino-acids, and the quality or value of a protein food is judged today by its amino-acid make-up. Proteins contain many amino-acids. Some are nutritionally more important than others. For example, the so-called sulphur-bearing amino-acids, methionine and cystine, are extremely important in poultry nutrition.

During digestion the proteins are broken down into their constituent amino-acids. These are absorbed into the blood via the gut wall and resynthesized into body amino-acids by the bird. They are then used in maintenance, production and body growth.

The balance of amino-acids in the bird's ration is vitally important, because an excess of one can lead to a higher requirement of the others. Naturally, when this occurs, growth rate and rate of egg production are impaired. Excess amino-acids in the form of proteins are not only dangerous to the birds' health but they are also wasteful. Whilst a slight excess may be converted into carbohydrate and used as an energy supplier, a great excess is converted into uric acid by the kidneys and excreted.

The bird's ration must therefore supply the optimum amount of amino-acids according to the rate of growth, production and species. A mixture of protein-supplying ingredients is more satisfactory than just one ingredient. Protein requirement is highest when the chick is growing and decreases as rate of growth decreases. Once production commences the requirement increases again.

Carbohydrates

Carbohydrates are composed of carbon, hydrogen and oxygen, the latter being in the ratio of two to one, as in water.

Carbohydrates are energy suppliers which keep the bird's body warm and allow it to do work, i.e. produce eggs and meat. They are an essential part of modern rations and often form up to 75 per cent of the ration.

The principle ingredients which are used as suppliers of readily available energy are: maize meal, ground milo (sorghum) and ground wheat. All these meals have relatively low fibre contents, which is typical of the so-called high energy ingredients. Oats and barley, whilst they also supply energy, do so at a lower level because of their higher fibre values. Thus it can be seen that high energy and high fibre values do not go together. The amount of energy to be included in a ration will depend principally upon the type of stock being fed and the food capacity of the bird.

Small-bodied laying birds, for example, will have the same requirements for energy as a large-bodied laying bird when based on each one pound of body liveweight. However, as the smaller bird will consume less food than the larger bird, the energy level of the ration should be proportionately higher to compensate for the lower food consumption.

The use of so-called 'high-energy' foods for the production of eggs and broilers is now commonplace in the poultry industry. It must be mentioned, however, that large-bodied laying birds, because of their higher food consumption, should not be fed laying rations which provide excess energy; this will cause overfatness and reduced egg production.

Fats and oils

Fats and oils are also energy suppliers. They belong to a group of chemical compounds called 'esters'. On digestion the fats and oils are broken down in fatty acids and glycerine. Fats and oils supply two and a quarter times the amount of energy that carbohydrates do. Thus, for every 1 pound of fat 2¼ pounds of carbohydrates are required to supply the same amount of energy. Because of this, fats are sometimes used in poultry rations in order to achieve a greater concentration of nutrients.

Energy when applied to poultry nutrition is generally expressed as metabolizable energy. This is the amount of energy which remains after a food has been digested and excreted. The amount of energy in a pound of foodstuff is often expressed in calories per pound or kilo calories per pound. The metabolizable energy values of the common ingredients used in poultry foods are well known and are therefore listed in the table dealing with 'Composition of feedingstuffs' on page 101.

Water

The removal of water from the diet of a bird will cause its death in a very short period. Food, on the other hand, can be withdrawn for a much longer period without causing death. It is interesting to note, in this respect, that the egg contains some 65–68 per cent water and the body some 60 per cent. Thus the importance of a constant supply cannot be overemphasized. This, of course, not only applies to laying poultry but to all other classes of stock and species of fowl.

Water helps in maintaining a steady body temperature. It is a major constituent of blood. It is necessary in the removal of waste body materials. One hundred adult laying poultry consume about 5–6 gallons of water in a day. In summer, during hot weather, this may increase by 25 per cent. Water consumption figures may be calculated by knowing that for each pound of food eaten 2 pounds of water will be drunk.

Water troughs must be cleaned out regularly to eliminate disease risks. Water should never be allowed to freeze, and outside water containers should be emptied each night during frosty weather.

Minerals

Minerals are just as essential in the ration as other ingredients. The bird's skeleton is composed mainly of calcium and phosphorous, and it is therefore essential that the diet contains the correct amounts of these minerals. The egg shell consists principally of calcium carbonate. A 2-ounce egg will contain about 2 grammes of calcium. Thus, for laying birds, the calcium requirement is higher compared with non-egg-laying stock. The work of both calcium and phosphorous is closely related to the vitamin D content of the ration. A lack of calcium, phosphorous or vitamin D can lead to serious breakdown in both the laying and the growing bird.

Salt is an essential mineral in all poultry rations. It should not be provided in excess, for this will lead to abnormally high water consumption. The minerals manganese, zinc, iodine, and iron are all required in small amounts. Their presence at the correct levels is essential for growth production and reproduction.

Most poultry rations supply part of the bird's mineral requirements, but under modern housing conditions and with high performance stock it is necessary to add small amounts of synthetic minerals.

Vitamins

In comparatively recent years research workers have found that, in addition to proteins, carbohydrates, fats and minerals, certain other substances are necessary to obtain good production and good liveability. These substances are called vitamins. Lack of vitamins can lead to what are known as deficiency diseases, particularly where the birds are kept under intensive management conditions.

Vitamins are classified as being either fat or water soluble. The fat soluble vitamins are: vitamin A, D_3 and E, whilst the water soluble ones are: vitamin B complex, vitamin K and C.

Fat soluble

Vitamin A

This vitamin is necessary in small amounts to maintain health. Lack of vitamin A causes keratinization of the mucous membranes, such as the eyes. Deficiencies in laying and breeding stock result in a reduced production, susceptibility to parasite (worms) infestations and a reduction in hatchability. Grass meal, maize meal and fish liver oils are rich natural sources of vitamin A.

The vitamin A in both yellow maize and grass meal is called carotene, which the bird converts into true vitamin A. Fish liver oils are not commonly used today because of the risk of rancidity. Synthetic sources in a dry form are now used in modern poultry rations.

Vitamin D

The bird requires a continuous supply of vitamin D_3 when housed intensively and kept away from sunlight. Free range poultry can synthesize their own D_3 in the body because of the sun's rays. It is essential in mineral utilization and absorbtion. Without vitamin D_3, poultry, particularly young chicks, develop rickets. In laying birds, absence of D_3 leads to poor egg production, and thin- and soft-shelled eggs. Leg weakness may also occur in laying poultry, particularly with battery-housed birds.

The vitamins may be provided by fish liver oils or as a synthetic preparation. The latter method is preferred because of the risk of rancidity in the fish oil. Sunlight, which also produces vitamin D_3 in the bird's body, is of no use when it passes through ordinary glass. Glass prevents the important ultra-violet rays from passing through. Only poultry which are housed extensively during the summer months can benefit from the ultra-violet rays.

Vitamin E

This vitamin is found naturally in the germ of cereal grains, and green foods. Lack of vitamin E in breeding stock can cause poor hatchability. The embryos usually die at the third and fourth days of incubation. In chicks an absence of vitamin E causes the disease known as Crazy Chick Disease. The bird is unable to maintain its balance as the disease affects part of the brain tissue. Crazy chick disease usually occurs when the chicks are between three and eight weeks old. Vitamin E is affected by rancidity and for this reason it is now usual to incorporate an anti-oxidant into the ration to prevent the oxidation of fats which causes rancidity.

The vitamin must be added in a synthetic form to rations which are fed to breeding stock.

Water soluble vitamins

Vitamin B complex

The vitamin B complex embraces a large number of B-group vitamins. These are thiamin (B_1), riboflavin (B_2), pyridoxine (B_6), vitamin B_{12}, folic acid, pantothenic acid, nicotinic acid, inositol and choline. They are all required for health and efficient production of meat and eggs.

Ingredients such as yeast, dried skim milk and fish meal are rich sources. Rations for high producing, intensively housed stock are usually supplemented with synthetic B-group vitamins.

Vitamin K

This vitamin is known as the anti-haemorrhage vitamin. Lack of vitamin K delays the blood-clotting time. High performance rations are sometimes supplemented with vitamin K.

Vitamin C

This vitamin does not play an important part in poultry nutrition.

Antibiotics

The word antibiotic means against life; that is, the life of micro-organisms. Antibiotics are produced by the growth of micro-organisms. Many thousands are known but relatively few are used, and permitted to be used, in poultry rations.

Antibiotics are used for two main purposes. Firstly, they may be used at relatively low levels to stimulate growth rate and improve the efficiency at which the bird converts food into meat. The names of the antibiotics and the amount to be used are the subject of Government legislation and only those on the approved list are permitted. Secondly, they are used therapeutically at much higher levels to control certain diseases. In this case a veterinary surgeon has to prescribe the antibiotic.

At the present time legislation allows the use of the following antibiotics for growth promotion in broiler and turkey production: Zinc Bacitracin, Flavomycin and Virginiamycin. In addition, non-antibiotic growth promoters may alternatively or also be used, viz.: Payzone Nitrovin, Grofas and some Arsenicals. It is thought that both antibiotics and non-antibiotics increase growth rate and improve feed utilization by enhancing the absorption of food nutrients, such as proteins and their components, the amino-acids. For some, however, the action is none too clear. In broad terms there is little to choose between any of them when used singly, but certain combinations have been shown to give additive and even growth increases greater than the effect of the sum of the two used separately, that is to say a synergistic effect.

For therapeutic use the antibiotic selected must be shown to have a direct effect upon the organism responsible for disease.

'Stress', which means any deviation from the normal, can sometimes be combated by using high-level antibiotics. Some conditions of stress are: neck moulting, debeaking, vaccination, etc.

Unless exceptionally high levels are used in the foodstuffs fed to ducks and geese, very little benefit is obtained with these two species.

D

Calorie–protein ratios

The protein or amino-acids in a ration are used most efficiently when the ratio of protein (or each amino-acid) is correct. The ratio of protein to energy (calories) varies with different species, age of stock and performance potential of the bird.

The younger the stock, the narrower is the ratio of protein to energy. The calorie to protein ratio means that for each 1 per cent crude protein in a ration a given number of calories are required.

With broilers, for example, the ratio for birds up to the age of about five weeks is 58 calories for each 1 per cent protein. With laying hens the ratios are difficult to establish. As a guide, 80–84 calories are allowed for each 1 per cent protein. It cannot be stressed too much that calorie-protein ratios are used only as a guide in formulation. They are not exact requirements.

Food requirements of the fowl

The various rations fed to poultry differ according to the species, age and purpose for which the stock is kept, i.e. table birds or laying stock.

The chick

The chick grows at a very rapid rate, especially during the first eight weeks of its life. Its nutritional requirements are therefore extremely high, as it is vitally important to get the chicks off to a good start in life. This necessitates the use of a chick ration high in energy and with a narrow calorie-protein ratio. The protein content should be approximately 20 per cent and the energy, in terms of metabolizable energy, should be around 1250 calories per pound of food.

The following is a typical ration for feeding to the chick during the first six to eight weeks of its life. No grain should be fed with this type of ration.

CHICK RATION

Ingredients	Per cent
Maize meal	35

Ground oats	10
Wheat meal	15
Barley meal	15
Grass meal	5
Soya bean meal	7
Fish meal	9
Dried skim milk	1¼
Limestone flour	¾
Steamed bone flour	¾
Dairy salt	½
Synthetic vitamins (Vitamins A, D$_3$ B$_2$)	¾
	100

Minerals. Add to the above ration, per ton of food:
8 oz manganese sulphate.
4 oz zinc carbonate.

The growing bird

The growing ration should be introduced after six or eight weeks of age. It is fed, in the case of growing pullets, until the birds are sexually mature, i.e. about twenty weeks of age.

As the grower has a relatively low requirement for protein, a level of around 15 per cent is satisfactory.

The energy requirement of the grower depends to some extent on the rearing environment. Pullets which are reared under constant short daylights should be fed a ration which supplies about 1200 calories of metabolizable energy for each 1 pound of food.

The ration given below may be fed in conjunction with ½ an ounce of grain per growing bird per day. Calcium-supplying grits should not be given.

GROWER'S FOOD FOR INTENSIVELY HOUSED PULLETS

Ingredients	Per cent
Maize meal	20
Ground oats	19
Wheat meal	25

Barley meal	15
Grass meal	5
Soya bean meal	7
Fish meal	4
Limestone flour	2
Steamed bone flour	2
Dairy salt	$\frac{1}{2}$
Synthetic vitamins (Vitamins A, D_3, B_2)	$\frac{1}{2}$
	100

Minerals. Add to the above ration, per ton of food:
6 oz manganese sulphate.
3 oz zinc carbonate.

The laying bird

The laying ration should be introduced at about twenty weeks of age. There are deviations to this recommendation, particularly with later maturing pullets. Whatever happens, however, the laying bird should be introduced to the laying ration just before the commencement of egg production and not left until production has reached, say, 20–30 per cent.

The ration for the small-bodied hybrid laying pullet should be high in energy and protein, but for the heavier type of birds the levels should be lower to prevent the birds becoming overfat.

The protein requirement for laying pullets weighing about $4\frac{1}{2}$ pounds is 15 per cent and the energy level recommended is between 1200 and 1250 calories per pound of metabolizable energy.

The ration layed out below needs to be supplemented with soluble oyster shell or limestone grit at the rate of 8 pounds per 100 pullets per week.

LAYING RATION FOR LIGHT-HEAVY-TYPE PULLETS

Ingredients	Per cent
Maize meal	20

Wheat meal	54
Grass meal	5
Soya bean meal	10
Fish meal	7
Limestone flour	2
Steamed bone flour	1
Dairy salt	$\frac{1}{2}$
Synthetic vitamins (Vitamins A, D_3, B_2)	$\frac{1}{2}$
	100

Minerals. Add to the above ration, per ton of food:
> 6 oz manganese sulphate.
> 3 oz zinc carbonate.

The breeding hen

Essentially, the nutritional requirements of the breeding hen are similar to those of the high producing laying hen. The exceptions are for the vitamins and minerals, which, because the bird produces an egg which contains a living embryo, should be included at higher levels. In particular, members of the B complex and the vitamins A and D_3.

The quality of the protein should also be increased, whilst the energy requirement is no different from that of the laying hen.

BREEDING RATION

Ingredients	Per cent
Maize meal	15
Wheat meal	$24\frac{3}{4}$
Barley meal	18
Ground oats	10
Grass meal	5
Soya bean meal	$11\frac{1}{2}$
Fish meal	8
Dried distillers solubles	$3\frac{1}{2}$
Limestone flour	2
Steamed bone flour	1

Dairy salt $\frac{1}{2}$

Synthetic vitamins
(Vitamins A, D_3, B_2
and other members of
the vitamin B
complex) $\frac{3}{4}$

                  100

Minerals. Add to the above ration, per ton of food:
    8 oz. manganese sulphate.
    4 oz. zinc carbonate.

No more than $\frac{1}{2}$ an ounce of grain should be fed with ration to each bird per day. Both insoluble and soluble grits should be fed *ad lib*.

Turkey rations

The following are some recommended formulas for turkeys:

Ingredients	Ration and per cent inclusion rate			
	Starter	*Grower*	*Fattener*	*Breeder*
Maize meal	30	21	10	10
Wheat meal	15	20	45	$27\frac{1}{2}$
Ground oats	5	7	—	10
Barley meal	5	20	$25\frac{3}{4}$	10
Soya bean meal	19	11	8	10
Fish meal	15	9	4	8
Grass meal	5	5	—	5
Dried distillers solubles	$3\frac{3}{4}$	$2\frac{1}{2}$	—	5
Limestone flour	—	$1\frac{1}{2}$	3	2
Steamed bone flour	1	2	3	1
Salt	$\frac{1}{4}$	$\frac{1}{4}$	$\frac{1}{2}$	$\frac{1}{2}$
Vitamins*	1	$\frac{3}{4}$	$\frac{3}{4}$	1
	100	100	100	100

*Vitamin A, D_3, and B_2, plus additional E, K and B-group vitamins.

Minerals. To all the turkey rations add, per ton of food:
       8 oz of manganese sulphate.
       5 oz of zinc carbonate.

No grain should be fed with the starter and growing rations. A maximum of 2 ounces and 1 ounce should be fed with fattener and breeder rations respectively.

An anti-blackhead drug should be used in all these rations at the preventive level.

The starter food contains about 27 per cent protein and has a narrow calorie–protein ratio. The grower food has a lower protein content and a much wider calorie–protein ratio. The calorie–protein ratio of the fattener is widened even further by reducing the protein level to about 17 per cent.

This is done to encourage a good 'bloom' or finish on the marketed birds.

The starter is fed from day-old to five weeks of age, the grower five to ten weeks and the fattener from ten to sixteen weeks. Turkeys kept longer than sixteen weeks should be fed a second fattener from sixteen weeks until marketing age.

Rations for table duckling

Table duckling can be produced satisfactorily by feeding an ordinary chick ration from day-old until three or four weeks of age. From this time until they are marketed at about eight weeks, an ordinary table poultry fattening ration may be fed. For satisfactory weight gains the food should be fed either as a pellet or as wet mash. The pellets should be fed *ad lib.*, whilst the wet mash is given in four daily feeds. Breeding ducks will perform perfectly satisfactorily on a poultry breeder's ration. Where specialist duck rations are available, they should, of course, be fed according to the manufacturer's feeding instructions.

Notes on some of the common ingredients used in poultry foods

Alfalfa meal is dried lucerne and is a useful feed for laying stock. Yolks are improved in colour by its use. Its fibre content is high and this prevents its liberal use.

Barley is not extensively used in poultry rations. Where fed indiscriminately, trouble has been caused, but experimental work has shown that it may very well be used in a properly balanced ration. Barley meal is often used with success in fattening rations.

Beans, which are fed as meal, are rich in protein but deficient in fats.

Brewers' grains are rich in protein, highly digestible and a useful food for laying stock. Probably owing to their bulkiness and cost of transport, they are not used as much as they might be.

Fish meal is rich in high-quality protein and can be fed to all kinds of stock. Only extremely high levels taint eggs and flesh.

Maize is extensively used in poultry feeding, both in its kibbled state and as a meal. Essentially a carbohydrate food, it contains very little fibre. Yellow maize is used in nearly all cases. With layers, it has a definite influence on the colour of the yolk. White maize meal is used a great deal in fattening mashes, because the yellow variety has a tendency to increase the yellow colour in the fat of the bird.

Maize gluten is a by-product from the making of glucose and starch. It is rich in protein. *Maize germ meal* (a by-product in making corn flour) is rich in oil, and care must be excercised in its use as, in excess, it is bad for the birds.

Milk was once extensively used in rations for young stock, but its high price in recent years now precludes its use almost totally. Milk can be used, if economically available, either in its liquid form or dried and can be either a skimmed or separated milk powder. *Whey* resulting from cheese making does not possess the high protein content of other milk products, being chiefly a carbohydrate food, but is valued for its mineral and vitamin content.

Meat meal and *meat and bone meal* are rich in protein, although the quality is not as high as in fish meal. To a certain extent both products are used as a substitute for fish meal.

Oats and *Sussex ground oats*. Oats is a useful food, though not used now to the extent it was some years ago. Valuable for its mineral and oil properties, its high fibre content makes it less acceptable than wheat. *Sussex ground oats* have always been a stand-by in the fattening yard, and they are also used extensively in chick mashes. The whole oat, including the husk, is ground to meal in making *Sussex ground oats*. This is very different from crushed oats. When grinding, a little barley is often included with the oats.

Palm kernel meal can be given in a mash to replace some other cereal by-products, but it should be used with caution, owing to its unpalatability if fed in large quantities.

Peas, like beans, are fed as a meal. They are seldom used today in modern poultry rations.

Potatoes are a carbohydrate food. Chat potatoes have been used in mashes for poultry, but it is only recently that experiments have shown that they can be used to a much greater extent in a mash. As a war-time measure, steamed potatoes were used in a layer's ration with success to the extent of 70 per cent, the rest of the mash being made up of middlings and fish meal.

Rice is a carbohydrate food and is low in protein. Some years ago rice was used in chick feeds a good deal. It is highly digestible, and probably the cost is the reason for its limited use now. Care should be taken in purchasing *rice meal*. It consists of the bran and polish removed when the seed is being prepared for human consumption and, owing to the oil it contains, readily goes rancid on storage.

Soya bean meal (extracted) is a very valuable source of good-quality vegetable protein. It is extensively used in modern high energy rations for all classes of stock.

Wheat (including the by-products *bran* and *middlings*) is one of the most used grains in poultry feeding. It can be employed for all classes of stock. *Bran* is the outer covering of the wheat and is generally used to add bulkiness to the ration. It is useful both as a source of protein and as a source of vitamin B, but it has a high fibre content. Owing to its bulkiness, it is not used in large amount in chick mashes. *Middlings* (sometimes known as *seconds* or *thirds* in different parts of the country) is the residue in the milling of wheat flour and is extensively used in poultry mashes for all stock. Samples vary a good deal, there sometimes being very little difference between fine bran and middlings.

Dried yeast is fed to young chicks, growing, breeding and laying stock for its high protein and vitamin content.

Feeding methods

Feeding time is the best opportunity the poultry farmer gets to have a good look over the birds and see if anything is wrong, so time spent on feeding is never wasted. Where dry-mash feeding alone is practised, it will not be so easy, for the birds will be at the hoppers in twos and threes all day. With wet-mash feeding, or when corn is given, one can get practically all the birds close at hand.

There are several systems of feeding, the old way with wet mash, and the more modern practice of mash fed dry and grain feeding. It is possible to feed dry mash successfully without giving any grain whatsoever. Grain alone can also be fed. This is still the practice where birds are a side-line on the general farm, and have the run of the farmyard. To feed grain alone would not be economically sound on a poultry holding, nor would it, except in the spring flush, make for the maximum egg production.

Wet-mash feeding (not recommended for large poultry farms.) To prepare a wet mash does not mean taking a dry mash mixture and adding water to it. It is not so simple as that, if it is done properly, although it is a method often practised. Such ingredients in the ration as maize meal, Sussex ground oats, wheat meal and fish or meat meal should be steeped in hot water and left for a time. When thoroughly soaked, the mixture can be dried off by the middlings and bran in the mash. Bran should not be soaked. Care must be taken not to use too much water but only sufficient to be thoroughly absorbed by the different ingredients. Wet-mash feeding is a great advantage where only a few birds are kept or where a large institute or school is running a poultry department, for kitchen waste and surplus garden produce may with advantage be incorporated in a wet mash. When ready for use, the mash should be in a crumbly state, not in a sloppy condition. It should always be fed in troughs. Never scatter it on the ground where it can go mouldy and become a menace to the birds. One drawback to wet-mash feeding is the labour it entails on a large holding. The preparation, mixing and carting round the food every day does take up an amount of time which some might think could be more profitably employed in other directions. At the same time, there is probably not so much wastage from split food or possibly from vermin when wet-mash feeding is practised.

Dry-mash feeding is practised extensively. It has many advantages. The food can be mixed in bulk once or twice a week and carted round to the different houses. The saving of time and labour is obvious. Dry mash is fed in hoppers, which in the case of the larger houses are fitted inside the house, at such a height from the ground as to make it readily accessible to the birds. Since in most cases the hoppers are open all day, or most of the day, there is not that scramble which occurs amongst the birds in wet-mash feeding, and they all have ample time for satisfying their appetites. Dry-mash feeding can be very wasteful if a proper kind of hopper is not used.

The custom in some quarters of leaving the hoppers open all night is not to be recommended as it attracts vermin, and rats can make a big inroad into the mash if given the opportunity. For the same reason, galvanized hoppers are better than wooden ones. It has been said that dry-mash feeding is not so satisfactory and does not give such good results as wet-mash feeding. Experimental work does not bear out this contention, except in cases where the mash, when fed dry, proves unpalatable. In such cases, wetting the mash may result in an improvement. Equally good results have been obtained with both systems, and the one to use is merely a matter for the poultry farmer to decide for himself. The personal element in feeding counts for a good deal and, however excellent the food, good results will not be achieved unless the birds are properly fed. It is to be deprecated that more care is not taken with the feeding nowadays. To throw the grain down all in a heap is not a good practice. A certain amount of grain feeding is advantageous, for not only does it provide variety in the diet but it is also easy to feed, and keeps the gizzard properly employed. If not given until just before it gets dark, then it is better to feed it in troughs. If fed in the morning, it should be scattered in the litter, especially in large intensive or semi-intensive houses. This will get the birds scratching and actively employed. For the same reason, it can be broadcast on range after the birds are let out, if the weather is suitable.

Where balancer mash and grain feeding are practised, it is usual to feed 2 ounces of grain per bird per day, on the assumption that an adult bird will consume approximately 4 ounces of food per day, and it is desired to give 2 ounces of mash and 2 ounces of grain. A hard and fast rule should not be made in this direction, for it is a good thing to let the birds have all they will eat of a well-balanced ration. However, let half of it be grain. Wet mash may be given in the morning and grain at night, or the order may be reversed. It used to be contended that grain remained in the crop longer than the mash, and that was the reason for the custom of feeding the grain later in the afternoon. Experiment has proved the contention to be incorrect. When feeding corn in conjunction with dry mash, it is a good thing to give, say $\frac{1}{2}$ an ounce of grain in the litter in the morning per bird. The dry-mash hoppers should be left open all day, giving the balance of $1\frac{1}{2}$ ounces of grain as the last feed, in the trough for preference.

There is no one correct system to follow. The keen poultryman will adopt a certain way of feeding and watch results. If he feels that he can do better by changing his methods he will be wise to make the alteration.

Pellet feeding (that is, dry mash made up in the form of pellets) is being followed by many poultry keepers who use the battery system. An advantage in using pellets is that there is little waste in feeding. Where pellets or crumbs are

AVERAGE COMPOSITION OF SOME OF THE MOST COMMON FEEDINGSTUFFS

Feedingstuff	Moisture %	Crude protein %	Ether extract % (oil)	Crude fibre %	Carbo-hydrates %	Ash % (minerals)	Metabolizable energy in calories/lb food
Barley meal	13.0	11.14	2.09	4.86	66.01	2.90	1257
Maize meal	13.0	9.20	3.53	3.01	69.76	1.50	1537
Maize germ meal	7.4	20.20	7.60	9.10	52.80	2.90	1097
Oats—ground	13.00	11.14	4.22	9.27	58.77	3.60	1100
Rye—ground	11.20	11.80	1.60	2.31	71.13	1.86	1322
Sorghum	11.00	10.19	2.70	2.78	71.12	2.20	1424
Wheat	13.00	11.27	1.51	2.44	69.78	2.00	1390
Wheat midds	14.00	16.06	3.25	7.49	56.40	2.80	1069
Groundnut ext.	8.50	47.30	0.88	12.82	26.00	4.50	1228
Groundnut exp.	8.00	46.78	6.39	7.39	25.74	5.70	1276
Soya bean meal ext.	11.00	46.50	1.70	5.49	29.51	5.80	1189
Sunflower meal	10.00	37.57	1.75	18.85	25.33	6.50	858
Palm kernel meal	11.00	18.58	5.64	12.06	48.82	3.90	754
Fish meal	10.00	66.00	4.00	—	—	20.00	1230
Meat and bone meal	6.00	50.60	5.00	2.20	2.60	33.60	1150
Dried skim milk	6.10	33.50	0.90	0.20	51.70	7.60	1230
Whey—dried	6.50	13.10	0.80	0.20	69.70	9.70	1180
Grass meal	10.00	15.38	3.00	20.90	39.92	10.80	584
Molasses (beet)	23.30	6.70	0.20	—	61.60	8.20	—

fed to intensively housed stock, the intensity of light should not be high or cannibalism may result. Both systems of feeding are extremely popular, however.

No instruction will be given here on ration formulation, for although this was practical in the past with a limited knowledge of poultry nutrition, today a vast amount of scientific know-how is necessary in order that a modern high producing bird is able to perform to its full genetic ability.

Suffice it to say here that, whilst it is necessary to have a proper calorie–protein ratio, it is important that a multitude of other factors should be taken into account. It is not difficult to formulate a ration which is balanced yet is quite unsuitable for poultry. It is useless, for example, to give poultry foods which they cannot easily digest.

The bird is able to digest very little fibre. Therefore, the fibre level should not be more than about 5 per cent of the ration.

The function of the gizzard is to reduce the food to such a state that the digestive juices can set upon it. Feeding insoluble grit, such as flint and granite, is a great aid to the efficiency of grinding.

The quantities which should be fed vary with age of stock and environment conditions. It is important that the grit is not too small or this may cause a digestive disturbance.

Soluble and insoluble grits

Some laying rations contain sufficient of the mineral calcium for high rates of egg production. Others do not.

In the former case, it is usually only necessary to provide oystershell or limestone grit where the production rate exceeds 80 per cent.

In the latter case, these soluble grits should be fed at all production rates. Birds housed intensively on deep litter should receive them *ad lib.*, whilst battery birds should have the amount restricted to around 7-10 pounds per 100 laying birds per week. The grit should be sprinkled evenly on top of the birds' food.

Calcium grits should on no account be used to supplement

rations for chicks, growing stock or fattening birds, unless
specifically advised by the feedingstuff compounder.

As has already been mentioned, insoluble grits such as
flint and granite aid in food utilization. They also prevent
the gizzard from becoming impacted with grass or other
fibrous materials. Insoluble grits may be purchased in sizes
varying from a pin-head chick size to a marble size turkey
grit. The correct size for the stock is essential. Baby chicks
need only take amounts of the small pin-head size from four
days of age. A grower's size may be introduced at six to
seven weeks of age. It should be fed at the rate of $\frac{1}{2}$ a
pound per 100 birds once per month. With adult laying
stock, 1 pound per 100 birds per month of the adult size
grit is sufficient.

Food costs

The cost of a ton of feedingstuff should not be the sole
criterion of purchase. The poultry farmer should take into
account the efficiency with which the bird converts the food
into meat or eggs. In this respect the food cost per dozen
eggs or per 1 pound of liveweight is more important.
There does not appear to be any one brand of balanced
feedingstuff which is better than all the others. Sometimes,
however, one particular brand appears to suit a particular
type of bird or environment better than another.

Least cost rations

Most commercially available poultry foods today are
formulated with the help of a computer. The procedure used
involves recording the cost of each raw material to be
considered, its nutrient composition and the nutritional
requirements of the stock. The resultant ration is frequently
termed a least cost one because the combination of ingredients selected has produced the lowest possible cost relative
to the requirements of the bird. Such a ration, however, may
not necessarily be the optimum for the production of eggs
or meat at least cost. Thus, the formulation of 'optimum'
cost rations, although not the lowest cost per ton, is the
most profitable.

The reader should understand that computer-formulated rations, when produced by a qualified nutritionist, are in no way inferior to manually-formulated rations, but that they do enable the poultryman to produce eggs or meat at a lower cost.

Chapter 9

Table Poultry

Classification of various table poultry. Broilers. Killing and processing. Plucking. Cooling. Packaging. Marketing. Trussing poultry.

Poultry meat can be produced in a very short time compared with other kinds of meat. During the pre-war years birds specially reared for the table were regarded as a great luxury. The change today has been brought about by the rise in world living standards, which has increased the demand for all meat products.

However, poultry meat still forms a very small proportion of most people's total annual meat consumption. It is mentioned that a higher consumption is important in helping to lower the cost of poultry to the consumer.

Table poultry embraces the following: the boiling fowl, large roasting chicken, caponized poultry and broilers.

Boiling fowl

Boiling fowls are definitely a by-product of the egg-producing industry, for laying stock, having completed their productive egg-laying life, are placed on the market for use in the catering trade and to a limited extent by housewives in domestic cooking. The weight range is, of course, considerable, depending on the type of egg-laying bird discarded. The small hybrid produces a carcass of only 4 pounds, whilst the large heavy-type cross produces a 6- to 7-pound carcass. At the present time, the return for the boiling fowl is not very lucrative, mainly because such large numbers are marketed, thus more than satisfying the normal demand. Market discrimination is most noticeable against black feathered, moulting, yellow-legged, thin or over-

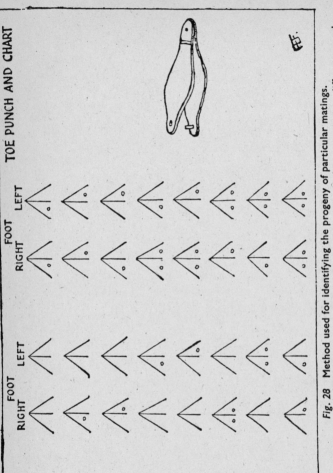

Fig. 28 Method used for identifying the progeny of particular matings. Toe punching is a valuable method of chick identification. In all there are sixteen different punch combinations, eight on each foot. The skin web between the toes is used for punching the holes.

fat hens and is shown by the lower prices received for them.

Colour preferences

In the past skin colour was a most important characteristic of table chicken; it is equally important today. In the U.S.A. and other countries, yellow-skinned varieties are popular. In the British Isles white-skinned birds are preferred. There is, however, no nutritional value in the colour, and the insistence on white- or yellow-skinned birds is a handicap to the trade.

A real table bird

To the trade a real table bird is one that comes from a breed well known for its good qualities, i.e. one that carries the maximum of breast meat. After being reared and brought along to about sixteen weeks, it has then been subjected to a term of fattening for two or three weeks prior to being killed and marketed. Twenty-five years ago this work was carried out almost exclusively in the Heathfield district of Sussex by a small body of men. The finishing process was a very skilled job, the men rearing their own stock. Today this sort of bird, called the 'Surrey', is rarely produced because of the high labour requirement and relatively slow turnover.

Large roasting fowl

Large roasting chickens are usually produced from cockerels which are the by-products of the pullet chick industry. Normally the heavy-type crosses are chosen because of their greater maximum weight and better carcass finish. The birds are generally sixteen to twenty weeks of age when killed and weigh about 8 pounds liveweight, although this will depend upon the strain or cross of bird used. The greatest demand for this type of bird is usually at Christmas and other festive occasions.

The popular crosses used in large roasting chicken pro-

duction are: Light Sussex crossed with a Rhode Island Red,
Light Sussex crossed with a New Hampshire Red and Light
Sussex crossed with an Indian Game. The male progeny of
this last cross is often crossed back onto the Light Sussex
female. Many general farmers produce roasting fowl because
the birds can be turned out onto the stubbles after harvest.
In this way they earn part of their own keep.

The small producer will find a ready farm-gate sale,
especially if the birds have been well fattened and are
presented in an attractive manner.

Caponized poultry

A capon is a male bird from which the testes have been
completely removed by surgical operation. It bears the same
relation to a cock as a steer does to a bull, or a gelding to a
stallion. The advantages of the capon over the 'cock' as a
table bird are a greater final weight, a more tender flesh
and a better finished carcass. The operation is not simple
to perform, and because there is an element of danger to the
bird's life it is not practised much today.

With today's demand for a smaller-bodied bird, the
technique of chemical caponization has become very popular. It involves the insertion of a small pellet under the skin
at the base of the bird's head or in the neck. The pellet is
a preparation of the synthetic female hormone known as
oestrogen. The hormone causes the testes to regress both in
size and activity for a period of approximately six to seven
weeks. After this length of time, the effect will slowly wear
off and the testes commence to function again. However,
during the caponization period, the caponized male deposits
a thin fat layer just under the skin surface, called sub-
cutaneous fat. This fat layer imparts a 'creamy-like' appearance to the carcass.

The best age for treatment is approximately five weeks
before the intended slaughter date. No bird should be
caponized under eight weeks of age as the hormone interferes with bone development. The hormone pellet is injected
under the skin with a small 'gun'. The needle of the gun
must on no account penetrate the flesh.

Each pellet contains 15 mg of oestrogen. Normally it is necessary to inject only one pellet, but if the birds are more than twenty weeks old two may be given simultaneously.

Some synthetic oestrogens can be included in the food. This method eliminates the necessity of catching and handling each bird, although the costs are proportionately higher. Oral administration should be discontinued two weeks before killing.

The capon usually demands a small price premium over the uncaponized birds which more than compensates for the additional handling and material costs. Popular capon weights are 6–8 pounds liveweight. Modern mass brooding techniques are often used to produce the capon in a similar manner to the broiler. When reared intensively, 2 square feet of floor space should be allocated per bird.

Broilers

The term broiler denotes an intensively reared table chicken weighing between $2\frac{3}{4}$ to $4\frac{1}{2}$ pounds liveweight and killed at eight to ten weeks of age. Today almost 300 million broilers are produced annually in the United Kingdom compared with less than half this number eight years ago. The broiler industry is extremely specialized, requiring not only an excellent knowledge of mass-production techniques but also big business know-how. The number of birds each producer rears is generally governed by the packing station which handles, processes and markets the produce.

Nearly all the strains and breeds used in this country are the result of American inported breeding stock. Their ancestry is based on the following breeds: White Plymouth Rock, Cornish White and New Hampshire Red. Complex breeding systems have been used in producing the modern broiler chick.

Housing

Houses in which the ventilation, light and temperature are all automatically controlled are now commonly used.

The birds are floor litter reared in batches of 5000 to 20 000.

Dim lighting is used not only to control cannibalistic tendencies but also to keep the birds in a subdued state. In this way they will grow faster and convert food into flesh more efficiently.

Each bird is allowed about half a square foot (76 square inches) of floor space from day old until it is marketed.

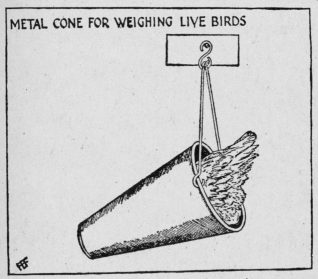

METAL CONE FOR WEIGHING LIVE BIRDS

Fig. 29   This saves a lot of time when weighing.

They are fed on carefully, scientifically-formulated rations which enable them to convert $2\frac{1}{2}$ pounds of food into 1 pound of liveweight in about nine weeks. Drugs to control such diseases as coccidiosis are always included in broiler rations. Rations which are high in energy, and therefore low in fibre, are used. These are provided in a crumb and pellet form.

After each batch of broilers has been sold, the house is disinfected and prepared for the next batch. Thorough disinfection is vitally important in the control of parasitic and respiratory diseases.

Profit margins are extremely small, and newcomers to the industry are not advised to participate in broiler production.

Like the broiler-growing industry, the broiler-breeding section is a specialized business, and suffice it to say here that it requires considerable knowledge to make it a paying proposition.

Killing and processing

It is customary to starve poultry for twelve to sixteen hours before killing, the shorter period being recommended for broilers. Water should be made available to all birds which are being starved.

The most common method of killing is by dislocation of the neck. The bird is held downwards, feet and shanks in one hand. Sometimes the tips of the wings are also held in the same hand to prevent the bird from flapping. The head is grasped in the other hand, with the first and second fingers forming a fork behind the bird's skull. The head is then forced backwards and downwards. The neck should break behind the head leaving a cavity of 1–2 inches, into which the body blood will drain. Killing by bleeding is also popular in large packing and killing stations. The bird is stunned first and then the jugular vein is slit with a sharp knife. Bleeding gives a whiter carcass than dislocation.

Plucking

The bird should be plucked whilst the carcass is still warm.

Dry plucking by hand is practised by small poultry keepers. It is best carried out in a sitting position with the bird's head hanging downwards. The flight and tail feathers are removed first. The breast feathers should be plucked next, followed by the sides, back, legs and wings. Care should be taken in plucking the breast feathers, especially over the narrow feather tract areas.

After plucking, all 'stub' feathers must be removed with the aid of a blunt knife. When 'stubbing' has been completed, the bowels should be emptied by squeezing the abdomen. To finish the job wash the shanks and singe off the body hairs with a methylated spirit or gas flame. Dry plucking may be carried out by machine. The feathers are removed by revolving plates.

Wet plucking is extremely popular in the large poultry packing stations. The birds are dipped into hot water (120°–130°F) for fifteen to thirty seconds, depending on the weight of the bird. Feathers may then be stripped by hand or by rubber fingers mounted on a rapidly revolving drum.

Cooling

Carcasses should be cooled as soon as possible after plucking, at a temperature below 50°F. On the poultry farm a cool room or cellar may be of considerable help. The use of slush ice and blast freezing achieves rapid cooling in large packing stations.

Packaging

The use of lightweight non-returnable boxes of wood or cardboard is today common practice. The containers are lined with grease-proof paper and contain six to twelve birds. The carcasses are arranged breast uppermost and covered with grease-proof paper.

Wrapping birds individually in cellophane is now very popular. Vacuum packing is also popular. Common wrapping films used are cellulose film and polythene.

Marketing

Owing to the competitive attraction of other types of meat, the future of the table poultry industry depends on the presentation to the housewife of a standardized and high quality article.

Packing stations

A high proportion of the table chicken output is presented

to the wholesale trade by packing stations. The birds are usually purchased on a liveweight basis. This type of marketing is used extensively in the broiler industry. It cannot be overemphasized that producers growing table chickens must first obtain a market. Success will depend on this.

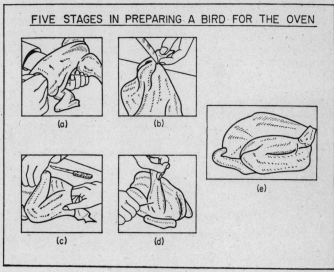

FIVE STAGES IN PREPARING A BIRD FOR THE OVEN

(a)   (b)   (e)

(c)   (d)

Fig. 30

Marketing direct to caterers and shops

There is considerable opportunity in this type of outlet. Grading should be carefully carried out and the birds must be attractively packaged. Small producers will doubtless supply only fresh birds, whilst the larger table poultry producer may supply oven-ready frozen birds.

Sales direct to the consumer

This is the most positive method of marketing. Only first-quality birds should be sold. They must not be badly plucked and presented or the customer may turn elsewhere for his poultry meat. Placing the birds in polythene bags is a more hygienic method of selling than selling 'in the flesh'.

Trussing poultry

In private trade it is very necessary to be able to truss poultry attractively, since a nicely trussed fowl commands more money than one trussed badly.

When trussing for roasting, take the plucked bird and lay it breast downwards on a table. Make a slit across the back of the neck and expose the neck bone. Cut around this bone close to the bird's 'shoulders'. Break off the neck at this point and remove the bone from the bird's head. Now pull away the bird's crop. Pass the forefinger into the body and loosen the lungs and other internal organs (see Figure 30 (a)).

Turn the bird onto its back, and make a cut between the vent and tail piece (parson's nose) (b). Insert two fingers and loosen the internal organs. Next withdraw the gizzard and all internal organs (c). Cut off the intestines at the vent and also cut out the vent. Wipe out the bird with a damp, not wet, clean cloth.

Now take an 8–10 inches long trussing needle and some thin clean white string. Place the bird on its back and press its thighs level with the table. Thread the needle and pass it through the folds of the legs and out the other side (d). Turn the bird onto its breast and thread the needle through the closed wings. Pass the needle in the opposite direction through the other wing. Untie the threaded needle and tie the two end pieces of string together.

Re-thread the needle and pass it through the end of breast muscle. Unthread the needle and cross the two ends of string round the parson's nose. Tie to complete the trussing (e). Turkeys may be trussed in the same way as fowls.

Chapter 10

Turkeys, Ducks, Geese and Guinea Fowl

Breeding, incubation, brooding, rearing and fattening.

Unlike other classes of table poultry, possibly with the exception of geese, turkeys are associated in the mind of the public with Christmas. Despite this, the consumption of turkey meat has shown a considerable increase since the early post-war years. Since these times, expansion within the turkey industry has been rapid. Today, something like fifteen million birds are consumed annually. Over half of these are eaten other than at Christmas time.

Market requirements fall into two classes. Firstly, the traditional Christmas trade. The type of bird demanded at this time of the year is much heavier compared with turkeys marketed in the interim periods. Weights are usually from 12 pounds upwards. No upper limit is given because many much heavier birds are used by the catering trade.

Secondly, the smaller all-the-year-round turkey. The weights are much lower and the range is narrower. The most popular weights are 5–8 pounds (oven ready).

Breeds of turkeys

The most common breeds used are Broad Breasted Bronze, Beltsville White and British White. Whilst these three breeds constitute the main breeds used in British turkey production, breeding has produced hybrid turkeys involving crosses betweeen these three breeds. The trend is for a white, broad-breasted turkey; white because of the absence of dark stub feathers when plucked. The bronze and black varieties do not 'finish' so well until almost fully grown.

Breeding

Turkey hens are not prolific egg producers, and on average

WHITE AUSTRIAN TURKEY

BLACK NORFOLK TURKEY

Fig. 31 Two useful breeds.

they lay about 100 eggs per bird in each breeding season. The smaller types of turkey are more prolific; the heavier or larger types are less prolific.

Characteristics such as egg production and hatchability depend on family selection for improvement. This involves progeny testing. Records are obtained by trap-nesting the females.

The normal breeding season is between four and five months.

With the trend towards greater breast width and shorter shank length, turkey breeders have difficulty in obtaining good fertility. Because of this, many turkey breeders now use artificial insemination, either wholly or partly. On modern breeding farms, the hens are housed in cages and inseminated every fourteen days. With natural mating, one male is mated to every eight or ten turkey hens. Artificial insemination can considerably increase this number.

Housing

Turkey breeding stock may be housed in a variety of ways, one of the most popular being the fold unit which measures 20 feet by 5 feet wide. Straw or pole yards are popular also for breeding turkeys. These are simply constructed of wire netting, poles, corrugated asbestos and three-ply felt. About 8 square feet of floor space should be allowed each breeder and 10 inches of food trough space. Under normal circumstances the hens will commence laying in March, when they should be about twenty-eight weeks old.

Incubation

In the main, the principles of incubation outlined in Chapter 5 apply equally well to turkey hatching eggs. The incubation period is, however, twenty-eight days, not twenty-one days. The eggs should weigh between $2\frac{1}{2}$ and $3\frac{1}{4}$ ounces each. Incubate only fresh eggs of good shape and good texture. The temperature for incubation should be 103°F in hot air natural draught incubators, 104°F in hot

AMERICAN BRONZE TURKEY

Fig. 32   Produces large birds.

water incubators and 99½°–100°F in large cabinet-type machines. The eggs should be tested for fertility on the twenty-fourth day in cabinet machines, and on the eighth or fourteenth day with natural draught machines. Hatching will commence on the twenty-sixth and twenty-seventh days, and should be complete by the twenty-eighth day. Depending on the fertility, the hatchability of all eggs set should be about 70 per cent. The hatch of the fertile eggs should be about 90 per cent.

Brooding and rearing the poults

The normal rearing period is eight weeks, although, depending on weather conditions, it may be reduced or lengthened. Whilst the brooders discussed in the Chapter dealing with the growing chick are suitable for turkeys, it must be remembered that only 50 per cent of the chick capacity should be used. On specialist turkey farms, intensive floor brooding is very popular. Batches of 1000 up to 5000 are brooded satisfactorily under one roof on similar production lines as those used for chicken-broiler production.

The following amounts of floor space are recommended for intensive floor rearing:

Age (weeks)	Floor space (square feet)
0– 4	¾
4– 8	1½
8–12	2½
12–16	3
16–20	4
20–24	5
24–28	6
28 ON	8–10

In the production of small-type poults, i.e. birds marketed between ten and sixteen weeks of age, specialist housing is necessary for maximum body weight gain and feed efficiency. The insulated broiler house is ideal in the production of this type of turkey. Both food and water trough space should be adequate so that all the birds can feed at the same time.

Turkeys for the Christmas market may be reared on good

grass in small runs or given free ranging. The grass should be kept short and the runs moved regularly to prevent a build-up of manure. It should be remembered that turkey poults need more room than chickens at all stages of their growth.

Feeding

For the first five weeks of the poults' life, the birds should be fed a turkey starter mash or crumb. This should be given *ad lib*. Adequate artificial light should be provided over the food troughs to encourage the poults to feed regularly and with ease. After five weeks a rearer ration should be introduced, and a little grain such as wheat and maize may be introduced. This type of ration should be fed to ten weeks of age.

In the past, turkey rearing was considered difficult. This was mainly due to lack of nutritional knowledge. The disease blackhead, which may attack turkeys of any age, is now completely controlled by the use of anti-blackhead drugs provided in the food.

Chickens and turkeys should not be kept in close proximity with each other because chickens harbour a worm which is the carrier of the blackhead parasite.

Turkeys should be marketed when they have acquired a suitable 'bloom' or 'finish' in their carcass. This may be helped by caponizing the turkeys five or six weeks before marketing with one of the oestrogen preparations discussed in the Chapter dealing with table chickens.

Ducks

Duck keeping today falls into two separate categories: the birds are bred either for table or for egg production. A great advance has been made in recent years in breeding for laying purposes, although the demand for duck eggs is small in comparison with hen eggs. The main objection to duck eggs is their comparatively strong taste.

This, however, is largely a matter of fancy. Some of the modern breeds are wonderful layers, often producing more eggs than laying pullets. Flock averages of over 300 are now reasonably common. The main breeds used in egg production are the Khaki Cambell and the Indian Runner. As ducks' excreta is extremely wet, the intensive systems of housing are unsuitable and free range is generally considered as the best method. Swimming water, whilst being an advantage to breeding ducks, is not essential.

Table duckling

This aspect of duck keeping is a highly specialized section of the poultry industry. The market is limited and the demand is only for first-quality white-skinned duckling weighing between $5\frac{1}{2}$ and $7\frac{1}{2}$ pounds. The duckling should be light in bone, and in this respect the Aylesbury duck fits the bill perfectly. Its only disadvantage is its low egg yield. However, when crossed with the Pekin breed, egg production is greatly improved without sacrificing meat qualities to any great extent. Dark-feathered duckling are unsuitable in this trade because of the objection to dark stub feathers.

Table ducklings are marketed when eight to ten weeks of age. After this period they moult and grow new feathers, and cannot be marketed until sixteen weeks old.

Breeding stock

The mating ratio for Aylesbury ducks is one drake to every four ducks. For Pekin the ration is one to six. When flock mating is practised, six to seven drakes are allowed for every twenty-five or thirty ducks. As the season progresses the number of drakes may be reduced. The average number of day-olds produced annually from each breeding duck is about fifty-five for Aylesbury and sixty-five for Pekins.

Breeding stock are selected according to the growth rates of their progeny, and also feed efficiency and carcass quality.

As it takes three months from the time the fertile egg is incubated until the duckling is marketed at eight weeks, it is necessary to plan the breeding season well in advance.

Incubation

Duck eggs have an incubation period of twenty-eight

E

ROUEN DUCK

PEKIN DUCK

Fig. 33 Good table ducks.

days. The principles of incubation previously discussed are very similar. Natural hatching with broody hens or artificial incubation may be used. Where broody hens or still air incubators are used, the eggs should be turned three times each day A liberal allowance of water is essential during the last few days of incubation. Eggs should be tested for fertility on the tenth, fourteenth or twenty-first days of incubation.

Fig. 34   The old favourite for the table.

Rearing

The growth rate, with good feeding, is phenomenal.

Age	Weight
Day-old	2 oz
2 weeks	14 oz
4 weeks	$2\frac{1}{2}$ lb
6 weeks	$5\frac{1}{4}$ lb
8 weeks	$6\frac{1}{4}$ lb

In order that these weights are achieved, the brooding should be good. No artificial heating is necessary after three weeks of age. Chick-type brooders are suitable for table duckling, although the number of ducklings reared in each should be reduced by 25 per cent. One tier brooder 8 feet long by 3 feet wide will rear sixty ducklings to three weeks of age. The brooder temperature should be 95°F to 100°F for the first few days. This is gradually reduced and completely done away with by three weeks of age.

Ducklings may be reared in solid floor houses. Under these conditions the following amount of floor space should be made available to each bird.

Age (weeks)	Floor space (square feet)
0– 2	$\frac{1}{2}$
2– 4	1
4– 6	2
6– 8	3
8–10	4

In fox-free areas free range rearing is satisfactory in the spring and summer months. Wire netting 18 inches high is sufficient to confine the ducklings from three to eight or ten weeks of age. Housing requirements are small, and curved corrugated iron sheets erected to form a shelter are perfectly satisfactory.

Feeding

The table duck should be fed *ad lib.* on a correctly formulated starter crumb or pellet. At three weeks of age this may be changed to a finishing pellet.

Water must always be available and at such a depth that the birds can completely immerse their heads. Antibiotics are of no benefit to fattening duckling unless used to combat specific diseases.

In order that the beginner can estimate the approximate food consumption of table ducklings at various stages in their growth, the following table will act as a guide.

FOOD CONSUMPTION GUIDE (PER DUCK)

Age (weeks)	Cumulative food consumption (pounds food)
2	1.8

3	3.6
4	6.3
5	9.25
6	12.60
7	16.00
8	19.50
9	23.00
10	26.00

Killing

The duckling should be starved for twelve hours before killing, so as to empty the crop and intestines. Water should be available. Killing is best done by dislocation of the neck, although bleeding is satisfactory. Death is instantaneous in both cases. Pluck the birds whilst still warm. The feathers should be carefully stored as they are a valuable source of income at 40p per pound. The feet should be scrubbed. They should not be packed for market until properly cooled. The loss of weight caused by killing is 5–7 per cent, bleeding 3–5 per cent and evisceration 28–35 per cent. These figures are expressed as a percentage loss of the initial liveweight.

Geese

Geese are kept almost entirely for table purposes, the main trade being at Christmas time. Profit may be made from selling goose eggs for hatching or young goslings for fattening.

The prices received for goslings are rarely high, so attention must be paid to keeping costs of production as low as possible. As geeses are good grassland grazers, they should be allowed to fatten for most of their life on this natural herbage.

Breeds

The main table breeds used are: Embden, Toulouse, Roman and English Greys. The Embden is a useful large all-white goose. It usually lays between twenty-five and thirty

EMBDEN GOOSE   CHINESE GOOSE   TOULOUSE GOOSE

Fig. 35 Three good types.

eggs each season. The Embden weighs 20–30 pounds at mature body weight. It is often crossed with the Toulouse for table purposes.

The Toulouse is a smaller goose than the Embden. It is grey and white in colour. Egg production is about forty eggs per bird per year.

The Roman is a small white goose which lays about sixty-five eggs in a year. It produces a useful 12- to 14-pound carcass in about six to eight months.

The Chinese goose is a light type and not much used for meat production. It is more prolific than the forementioned breeds, but has the disadvantage of a dark coloured meat.

Breeding

Mated breeding geese are referred to as a 'set'. A 'set' is usually one gander to three or four geese. Young geese are better mated to a second-year gander to obtain good fertility. Young geese commence laying in February and cease by mid-June. They should be mated up well in advance of the breeding season. Hatching is carried out in incubators or under broody hens. One hen will incubate three to four goose eggs. The geese should be fed a breeder's food throughout the season. A little corn may be fed. This is best given in the drinking water. Whilst swimming water is not essential, it does help to improve fertility.

Goslings only need heat for the first three weeks. During this period a good-quality chick mash should be fed. After three weeks of age grain should be given, but once the goslings are on good grassland this may be discontinued. In order that the birds fatten well, they should be fed grain during the last four to six weeks before killing.

The sexes may be easily distinguished by manipulation of the genital organs as early as day old. However, it should preferably be carried out at three to four weeks of age, when 100 per cent accuracy can be obtained. The female genital organ will be concave and smooth, whilst that of the male will protrude as a fleshy 'pencil'.

In adult geese, sexing without examination is often difficult. The gander usually has a longer and stouter neck.

Guinea fowl

Guinea fowls are rarely kept on commercial poultry farms, although with a good market the profit is very high. Housing is not required, for they will prefer to roost in the trees. They make excellent watch-dogs because of the piercing call which they make when approached.

The most common colour is pearl, although there are white varieties. Both sexes are similar in colour and shape, and it is often difficult to distinguish between males and females. One male will mate with three to four females.

GUINEA FOWL

Fig. 36

Incubation takes twenty-eight days and is best done under a broody hen. Heating or brooding is only required for one month, after which the chicks are quite capable of fending for themselves. They should be fed on good-quality chick food, preferably as a wet mash, five times each day for the first week. This should be reduced to four or three meals a day.

Chapter 11

Markets and Marketing

Wholesale and retail markets. The best methods of marketing. Packing stations for eggs and poultry.

Good marketing is essential for a successful business. One must, of course, provide just what the public requires to purchase. But, having successfully surmounted all the hazards associated with production, our marketing has in the past failed to reach those standards which we know are so important to an expanding industry.

Packing stations

Packing stations for the reception of new-laid eggs have been established. Eggs from the farms are collected by vans and brought to the packing station. Here they are cleaned if necessary (though this should be done by the producer himself), candled for freshness, graded for quality and size, and then packed in the proper cases for shops, hotels, etc. The packing stations pay the producer the price received for the eggs less a charge per dozen to cover the overhead expenses of the station. There is much to be said in favour of these stations, which up to 1939 were on a voluntary basis. However eggs are sold, they should be marketed fresh and clean. No eggs from 'stolen' nests, i.e. eggs laid away from home in hedgerows and the like, should be packed with those of known age. There is a risk of them being bad, and a bad egg will naturally mean a dissatisfied and probably a lost customer. If the producer is serving his own private customers, he should grade his eggs for size in the same way as the packing station.

There are various methods of marketing table poultry. The man in a small way of business will naturally look for private customers to whom he will retail his birds. Where the producer has a large regular output throughout the year, the private customer is not necessarily the best proposition. However many customers the producer may have, he is bound to have some birds left on his hands at times, which he will have to sell as best he may, and he will be lucky if he is always able to place these in a shop at the price he desires. Large consignments can be, and are, marketed through the hotels and large catering establishments, but here again there is the danger of supplies not being wanted at certain times. The large wholesale markets are a very satisfactory outlet for produce. However the quantities vary, the birds are always taken without hesitation, and a cheque in payment is returned promptly in the course of a day or two.

Grading chickens

Chickens should, of course, be graded for market, that is they should be of as near a size as possible in each package. Especially should this be the case in the non-returnable package, and the reason for this is obvious. If the packages can always be depended on to contain exactly what the invoice says, and nothing else, then confidence is established between producer and distributor, and again between distributor and purchaser. As a result, a great saving in labour, time and temper is effected. The purchaser can buy on the invoice with confidence and need only look at one sample package, or not even bother about that.

Table birds are now sold under the following grades:

Double Poussin	15–24 oz
Asparagus Chicken	$1\frac{1}{2}$–2 lb
Spring Chicken	2 –$2\frac{1}{2}$ lb
Summer Chicken	$2\frac{1}{2}$–3 lb
Harvest Chicken	3 –$3\frac{1}{2}$ lb
Michaelmas Chicken	$3\frac{1}{2}$–4 lb
Family Chicken	4 –$4\frac{1}{2}$ lb
Party Chicken	$4\frac{1}{2}$–5 lb
Banquet Chicken	5 lb and over
Boiling Fowl, small	3 –4 lb

Boiling Fowl, medium 4 –6 lb
Boiling Fowl, large 6 lb
Duckling 3 –4 lb
Norfolk Duckling 4 –5 lb

The 3- to 4½-pound chicken is often bulked under the term 'broiler'.

Continental efficiency

Marketing has been reduced to a fine art on the Continent, where it is done under government supervision. Producers would do well to make it their business to visit the large wholesale markets. It is an object lesson and it will pay them to do so. The distributors, though busy men, are always ready to find the time to discuss the market, and to give advice on marketing and market requirements.

The willingness to take any varying quantity of chickens at unstated times, together with prompt payments for consignments, makes wholesale marketing much more attractive than the uncertainties of retail trading, despite the increased prices received from the latter.

All types of poultry farm will at some time or other have old hens to dispose of. Birds finished with as breeders and pullets that have ended their laying season, and whose record does not justify their retention for another year, as well as hens that have gone through two seasons and are not quite right for the breeding pens—all these will have to go.

Local markets

There is a fair market for this class of fowl, i.e. old hens, at certain times of the year. The birds must, however, be well fleshed. Thin and out-of-condition birds will make next to nothing. Inquiries should be made at the markets as to when the birds will be required.

Ducklings for the table must be marketed when under ten weeks old. Where there are not many to dispose of, probably it is better to sell to private customers or the local poultry shops. If large numbers are to be sold, then the markets and packing stations will have to be the channels of disposal.

Geese in the same way can be marketed privately if not in large numbers. Plucking geese has to be carefully done. As with ducks, it requires more care than plucking fowls, and for that reason many producers like to dispose of their birds alive if possible.

Turkeys are often bespoken by customers some time before Christmas, and a good many of them can be disposed of in this way to private customers. There is usually a big demand for turkeys of all sizes right up to Christmas Eve, after which it declines. Unsettled mild weather will cause prices to fluctuate rapidly, as will the arrival of heavy imported consignments. Local markets will take turkeys alive, but the wholesale markets will want them killed and plucked. If large numbers are to be disposed of, they should be carefully graded when packing for market.

A marketable sideline

Poultry manure is a marketable sideline from all types of poultry holdings. It is not out of place to mention it here. True, it does not make the ready sale it should, probably because little has been heard of it. As a fertilizer it should be especially useful to market gardeners and nurserymen. Where birds are on range the manure cannot be collected, but where they are in fattening pens it can be gathered from the floor. It should not be kept in the open, where it will rapidly deteriorate, but should be stored under cover. Some people prefer to mix it with earth in the proportion of two parts poultry manure to one of earth. If it is possible to spread it on trays in a shed to dry, so much the better. In addition to nitrogen, poultry manure contains phosphates and potash. A growing bird voids about 25 pounds in six months, and about thirty adult birds will make 1 ton of manure in a year.

Feathers are quite a profitable sideline on a table poultry plant and buyers will take them at so much a pound. White feathers always make a higher price than the coloured ones, so they should be kept apart. For purposes of sale, wing and tail feathers must not be put in with the others. Horticulturists will sometimes take the latter to rot down or plough in for manure. The feathers for sale should be kept clean, put into clean sacks and hung up out of the way of mice to dry. When a good quantity is ready for marketing, buyers will often call and collect the feathers, paying for them on the spot. It is estimated that the feathers from 100 average birds (apart from the wing and tail feathers) will amount to 25 pounds.

Various sales

One hardly considers the sale of hatching eggs, day-old chicks, and growing and adult stock as coming under the label of marketing. However, since holdings are in existence whose sole object is the sale of this class of stock, their disposal may well be considered here. Breeding establishments have their own conditions which govern the sale of hatching eggs; these are usually sold in dozens, twenty-five to fifty or 100 and over. It is the rule to replace infertiles once where dozens are sold, but where fifteen eggs to the setting are supplied no replacements are made. It is usual to ask for the return of the infertiles before replacement. Proper egg boxes should be used for the dispatch of hatching eggs, as in most cases they will be travelling by rail and the railway companies will not usually accept them if they are improperly or carelessly packed. Whilst eggs will travel well in boxes made for the purpose, some people use sawdust as packing to make doubly sure.

Day-old chicks

Day-old chicks are sold straight from the incubator by breeding farms and hatcheries. It is usual to offer to replace any dying on rail if they are returned at once to the vendor. These day-olds are also packed in special boxes, usually in two-dozen lots. The boxes are lined with hay in order to keep the chicks warm in transit.

The sale of day-olds and hatching eggs is confined in certain measure to the early spring months, when the weather may be treacherous, hence the care that should be taken in their marketing. Growing stock, on the other hand, will usually be disposed of in the late spring and summer months. Care must be taken in the sale of birds at eight weeks or three months, or even over, to see that they are not overcrowded in their crates. If they have far to travel, they should be fed and watered before dispatch.

Stock cockerels and pullets, or hens for breeding, are generally required for autumn. It is usual to send these on approval, a stipulation that they be returned in two or three days if not acceptable being made.

Sales of eggs to hatcheries

The increasing demand for day-old chicks, both for laying and for the table bird production, has greatly increased the size of hatcheries—those buying hatching eggs to incubate and sell the day-old chicks. There has also been an increase in the size of breeders who cater for the day-old chick trade.

A number of farms now mate up stock for the sale of eggs to these hatcheries to the benefit of both parties. Eggs are sold to hatcheries at so much per 100. A bonus is often given for hatchability above a certain standard.

Chapter 12

Hygiene, Sanitation and Diseases

Recognition of symptoms, carriers of disease, disease spread, disinfection. Diseases—specific and non-specific.

It is of vital importance that the poultryman should have some knowledge of the diseases that he may possibly run up against in the course of his work. He must also know what steps to take to counter an outbreak pending the arrival of professional help. Above all, he should be thoroughly conversant with all the preventive measures necessary to keep disease at bay.

Hygiene and sanitation

Disease control can only be achieved with a sound knowledge of husbandry and hygiene. The correct spacing of birds, adequate ventilation, the prompt recognition and isolation of ailing birds, and general sanitation of housing and equipment are just as important as a knowledge of modern drugs. Subnormal health resulting from parasitism, faulty nutrition and chronic infections interferes with production potential and takes an even greater toll of profits than actual disease.

Recognition of symptoms

This can only be satisfactorily carried out by sending a typically affected bird for post mortem examination.

The second most important point in hygiene is that birds recently purchased from other premises or returned from other premises should be isolated from the home flock for at least three weeks before mixing them with the other poultry.

Purchases of stock should always be made through reliable and reputable sources. Chicks should be purchased only from flocks where routine blood testing is carried out.

In order to prevent adult stock spreading disease to young birds, the two age groups should be reared in isolation from each other. Traffic between the old and young birds should be reduced to a minimum.

Carriers of disease

The carcasses of disease-infected birds should always be correctly disposed of. They should be either burnt or buried. To leave them lying about only attracts vermin, which could spread the disease to healthy birds. Vermin should be controlled, for not only can they transmit certain bacterial diseases but they also contaminate food and water.

Clean egg production is absolutely essential in the control of egg-borne diseases. Bacteria and dirt are able to pass through the pores in the egg shell and, in fact, the embryo. This is apart from the lower return the producer will receive from his egg packing station for dirty eggs.

The best method of cleaning eggs is to use a dry abrasive. The eggs, unless very badly stained or coated in dirt, should not be washed. Wiping with a damp cloth hastens the penetration of bacteria.

Eggs may be dipped into germicidal solution containing a suitable detergent. Clean egg production can be improved by regular nest box collections, at least three times each day, and seeing that the nest litter is always clean.

How diseases are spread

Some diseases such as Coccidiosis—which is a parasitic disease—are spread by contamination of litter materials and food which is contaminated with droppings and other discharges. Warmth and humidity favour the multiplication of such diseases. Grass runs should be kept short to allow the sun's rays to penetrate. Soil disinfection is never easy, and the best method is to dig the ground bi-annually and rest it for the longest economic period. Parasites such as fleas and lice multiply quickly in dirty, dark conditions.

Houses must therefore be clean and well ventilated. Perches and nest boxes are the most common sites for these external parasites. The walls, floors and ceilings of houses should be smooth to allow for dusting, washing and disinfection.

Disinfection

Before new stock are moved into a house, the latter should have been thoroughly cleaned and disinfected. As disinfectants lose their efficiency in the presence of organic material, such as dirt and droppings, it is essential that the house is first thoroughly cleaned. The empty house should be scrubbed with hot water containing a 4 per cent solution of washing soda. When dry, the house should be disinfected with a product approved by the Ministry of Agriculture, Fisheries and Food.

In deep litter houses, the litter material should be completely removed after each batch of birds has vacated the building.

Diseases can divided into two main groups: specific diseases and non-specific diseases.

Specific diseases

These are caused by agents such as bacteria, viruses, parasites, fungi.

Bacterial diseases

B.W.D. (Bacillary White Diarrhoea). This disease is also known as pullorum disease. The bacteria responsible for B.W.D. is called Salmonella pullorum. Chicks are most commonly affected during the first two weeks after hatching. Mortality will vary from 10 to 80 per cent unless treatment is commenced. The symptoms are lack of appetite, huddling under the brooder and listlessness. Sometimes a white diarrhoea is present.

B.W.D. is usually transmitted via the hatching egg which has been laid by an infected hen. The infected hatched chick will quickly infect other chicks either in the incubator or in the brooder. Some chicks will recover, but they will remain carriers of the disease and must not, therefore, be used for breeding.

Treatment for this disease is by the drug furazolidone (Neftin), which is included in the food for ten days.

B.W.D. can be controlled by blood testing the breeding stock. This blood test detects the carrier hens, and once these are found and eliminated the disease is controlled. Blood testing can be carried out on the poultry farm quite satisfactorily. A sample of blood is collected from each breeding bird and mixed with a specially prepared antigen. If the bird is a carrier of B.W.D., the antibodies in its blood will cause the 'blood mixture' to agglutinate into well-marked clumps. In the case of the non-infected bird, the blood-antigen mixture will remain unchanged.

After removing the carrier birds, the houses should be disinfected and the non-reacting breeding birds re-tested at monthly intervals until no further reactors are found.

Fowl typhoid

This disease is also caused by a Salmonella bacterium. Outbreaks in young stock are comparatively rare, the disease usually affecting adult stock. Infected birds show lack of appetite, yellow diarrhoea and paleness of the head. Treatment can be carried out by using furazolidone. Carriers can be detected by using the same blood test as described for B.W.D.

Chronic respiratory disease

Respiratory diseases are becoming more common and serious under modern intensive management systems. Affected birds show a nasal discharge, which tends to harden and block the sinuses of the head, causing distortion and swelling of the face.

Overcrowding, faulty nutrition and bad ventilation all predispose to this disease.

Treatment can be carried out with antibiotics given in the food or the water.

Viral diseases

Fowl pest is one of the most serious of the viral diseases. It is highly infectious, causing mortality and serious economic loss due to severe depression in egg production and growth rate. Fowl pest is controlled by vaccination, but no vaccine gives absolute protection. All poultry flocks should be vaccinated in order to provide protection for the owner and neighbouring poultry farmers. At the same time, poultry farmers must take every sensible hygiene precaution to help eliminate infection risk. Vaccines are of two types: (a) inactivated, whereby virus has been killed by chemicals but is still effective, and (b) live vaccines —prepared from mild strains of the live virus of which two types may be used, (i) Hitchner Bl (ii) La Sota.

There are certain rules to follow when vaccinating. Vaccines must be stored according to manufacturers' instructions. Only healthy birds can be vaccinated. Vaccination equipment should be sterilized each time a new flock is vaccinated. Birds must be handled gently and quietly at all times. Live vaccines must be thoroughly mixed according to manufacturers' instructions. There are four main application methods for vaccine usage:

FOWL PEST VACCINATION PROGRAMMES

1. Emergency programme—when disease is prevalent.
A. Broilers:

(a) Vaccinate at day old	Hitchner Bl by eye drop, beak dipping or spraying.
(b) Re-vaccinate at 18-21 days.	As above.
(c) Re-vaccinate at 35 days	Hitchner Bl or La Sota in water or by eye drop.

B. Layers and turkeys:

(a) Vaccinate at day old.	Hitchner Bl by eye drop, beak dipping or spraying.
(b) Re-vaccinate at 18–21 days	Hitchner Bl, eye drop or water.

   (c) Re-vaccinate at 35 days   Hitchner Bl or La Sota in water.
   (d) Re-vaccinate at 10–12 weeks As above.
   (e) Re-vaccinate at 16–18 weeks As above.
   (f) Re-vaccinate every 5 months As above.

2. Normal programme—when disease is not prevalent.
A. Broilers:
   (a) Vaccinate at 21 days   Hitchner Bl via water.
   (b) Re-vaccinate at 35 days   Hitchner Bl or La Sota via water.

B. Layers and turkeys:
   (a) Vaccinate at 21 days   Hitchner Bl by eye drop or drinking water.
   (b) Re-vaccinate at 8–10 weeks   Hitchner Bl or La Sota by eye drop or drinking water.
   (c) Re-vaccinate at 16–20 weeks Hitchner Bl or La Sota by eye drop or drinking water.
   (d) Re-vaccinate all adult fowls Every 5 months with Hitchner Bl or La Sota in drinking water.

Note: It is emphasised that these vaccination programmes are guides to disease control and if in doubt the poultry farmer should consult a veterinary surgeon.

Marek's disease

Marek's disease is caused by a virus in its classical form and affects the central nervous system. There is an acute form which causes death without symptoms of disease. Infection to Marek's occurs in the first few weeks of life, although signs of disease may be delayed for several months. The virus is spread by contact from infected birds to susceptible growing chicks.

   Birds can be vaccinated against the disease at one day old, usually in the hatchery. This can, however, be carried out on the farm.

It is vitally important to store and use the vaccine strictly according to the manufacturer's instructions. Vaccination does not guarantee freedom from disease, but it should considerably reduce the risks.

Lymphoid leucosis

Caused by viruses. The disease usually occurs in chickens of four months and older, and is characterised by massive liver enlargement (big liver disease). Other organs may also be affected. Lymphois Leucosis virus is excreted in the droppings and saliva of diseased birds. Carrier birds may be completely healthy. No method of treatment or vaccination is available and the disease must be restricted by suitable flock management, such as attending to young stock before adults and strict attention to hygiene between batches of birds.

Epidemic tremors

This disease occurs in chicks between one and four weeks of age, and occasionally in older birds. The chicks are unsteady on their feet and frequently fall over onto their sides. Tremor of the neck and head is frequently noticed. There is no cure.

Parasitic diseases

Coccidiosis. This disease is a common cause of death in young chicks. It is caused by a small microscopic parasite which lines the inside of the intestinal tract. It causes the blood vessels in the intestines to rupture. There are many species of coccidia. The type which attacks the bird's caeca is the most common. The caeca are small blind pouches near the end of the small intestine.

Affected chicks appear listless with ruffled feathers. Their droppings are often stained with blood.

A number of drugs can be used to prevent coccidiosis occurring. These are usually added to proprietory brands of chick food and fed continuously for the first eight weeks of the chick's life.

For curative purposes, the drug is best put into the drinking water. This is because food consumption invariably declines in the presence of disease.

Blackhead. This protozoon disease principally affects turkey poults. It is very contagious, affecting the bird's liver and intestines. Affected birds are listless, lose appetite and have a yellowish diarrhoea.

Treatment is usually carried out by using an anti-blackhead drug in either the food or the water. As the turkey does not develop an immunity to blackhead, the drug should be given continuously from about five weeks of age until the birds are marketed.

Hens can affect turkeys, and the two species should not therefore be mixed.

Worms

The large round worm is about $1\frac{1}{2}$ inches long and almost white in colour. Heavy infestations will cause poor growth and reduced egg production. The use of the drug Piperazine is advised for treatment.

The small caecal worm can act as a transporter of the protozoal disease blackhead. It can be treated with a piperazine compound.

Gape worms are rarely found in chicks, but outbreaks do occur in turkeys and game birds. The worm, which is 'Y' shaped and red in colour, adheres to the wall of the bird's windpipe, causing gasping and death by obstruction. The drug Thibendazole can be used to kill the worms and control the disease.

Fungal diseases

Aspergillosis. Sometimes referred to as brooder pneumonia, this disease is more commonly seen in young turkey poults than chicken. It is difficult to cure, although the antibiotic 'Nystatin' has successfully been used. Aspergillosis is present in damp, mouldy hay and litter materials. The poults have difficulty in breathing and frequently stretch their necks in an endeavour to draw breath.

Non-specific diseases

This group includes nutritional deficiencies, constitutional disturbances and conditions associated with husbandry. A few of each will be discussed.

Nutritional deficiencies

A deficiency of Vitamin A gives rise to what is called nutritional roup. The eyes show a watery discharge. This later becomes a white cheesy deposit, which frequently blocks the nasal passage. Absence of vitamin A also causes a keratinization of epithdial tissue and reduces hatchability in breeding birds.

Vitamin D. Lack of this vitamin causes rickets. Bones become soft and rubbery. Inadequate calcium and phosphorus can also cause this condition.

Members of the vitamin B complex are necessary for good growth and reproduction, and a lack of vitamin E can give rise to the condition called crazy chick disease or nutritional encephalmalacia.

Constitutional disturbances

Egg peritonitis can be caused by birds which carry the bacteria of B.W.D. and fowl typhoid.

Affected birds have a penguin-like gait and a swollen abdomen.

Prolapse may be caused by egg binding. This occurs when an egg is too large for the oviduct and forces it to protrude as the egg is laid. Cannibalism frequently results and affected birds should promptly be removed from the flock.

Vices

The most common vices are feather pecking and cannibalism. Overcrowding and insufficient food trough space predisposes to them.

Debeaking may be used to control tendencies to vice.

Chapter 13

Records and Production Costs

Keeping accurate records is an essential part of modern poultry keeping, whether the unit is large or small. The economic position in all sections of the poultry industry is such that profit margins are slowly declining, which further emphasizes the need for maintaining records of production. This chapter sets out to provide the interested reader with the type of day-to-day records which must be maintained. It also furnishes information of production costs typical of those found on the efficient poultry farm.

Records which have been well compiled should, on analysis, save the poultry farmer work and worry. Properly kept and analysed, they will tell him which of his poultry enterprises are most profitable and, more important still, which ones incur a loss or at best only break even.

Many poultry keepers regard record keeping as a chore which, if possible, should be avoided. Keeping records is, of course, a chore but one which takes comparatively little time. Recording such factors as food consumption, mortality, eggs laid, litter used and approximate time spent each day on each work aspect can tell the poultry keeper much, especially if they are maintained daily. Food consumption is the most important factor to record because fluctuations may be the warning light of disease and a reduced egg yield. Knowledge of such fluctuations often enables the poultryman to take preventive action or call in a more experienced poultryman. Egg records, when transferred to graph paper, will show the production curve and tell the poultryman whether his flock is laying profitably. It is no use maintaining the records if they are not going to be analysed.

Records should be kept correctly. Scraps of paper and cigarette boxes are almost worthless as they are liable to be lost or thrown away.

144

The recording charts should be hung in the house, out of the birds' reach. To encourage their use a pencil should be attached to each one. The figures should be legibly entered.

Apart from the important day-to-day management records, the poultryman must, of course, record all his costs and returns. Not only is it essential in business but it also enables the farmer to cost accurately each section of his poultry enterprise.

Production costs

Rearing replacement—laying pullets

The cost of rearing laying pullets can have a marked effect on the cost of egg production. Today, modern pullet rearing is a highly specialized business.

Whilst production costs will obviously vary from farm to farm, depending on the type of pullet being reared, the method of housing and the system of feeding, the following illustration should act as a guide.

COSTS PER BIRD

	Light-type pullet	Heavy-type pullet
	New pence	New pence
Day-old chick	18.0	18.0
Food	35.0	40.0
Fuel and lighting	1.6	1.6
Labour	6.0	6.0
Mortality	1.0	1.0
Depreciation and sundry costs	3.5	3.5
	65.1	70.1

Egg production

As feeding accounts for about 65 per cent of the total costs involved in egg production, any saving in food wastage will materially affect costs and, therefore, the profits. Egg production costs are measured in many ways, the most common of which is the total costs incurred in producing a dozen eggs. This can only be done where accurate records are maintained.

The following illustration shows the breakdown in costs of a pullet recorded over a forty-eight-week-laying period:

	New pence
Point of lay pullet	67.5
Food	200.0
Electricity	5.0
Labour	15.0
Depreciation	15.0
Interest on capital	12.5
Mortality	5.0
	320.0

In this illustration the forty-eight week food consumption has been 90 pounds. Mortality has occurred at the rate of 1 per cent per month.

Broiler production

Profit margins are not high in the broiler industry. The facts and figures presented will give the reader an idea of production costs.

	New pence
Day-old chick	6.0
Food	17.5
Litter	0.3
Heat and light	0.9
Medication	0.5
Labour and management	1.0
Miscellaneous (including	
turn-round costs)	1.0
	27.2

These results would be based on factual information similar to that given below.

Age birds sold	60 days
Floor space per broiler	¾ square foot
Mortality	3.5 per cent
Food consumption per bird	9.0 pounds
Feed conversion rate	2.2:1

Capon production

	13-*week* capons New pence	18-*week* capons New pence
Day-old chick	9.0	6.0
Food	40.0	50.0
Heat and light	1.0	1.2
Medication	0.2	0.2
Labour and management	6.0	8.0
Miscellaneous	1.0	1.0
	57.2	66.6

In the case of the thirteen-week costing, a broiler-type chick is generally used. This should weigh between 6 and 7 pounds by thirteen weeks and have consumed about 22 pounds of food. The eighteen-weeks-old capon is usually a cockerel reared as a by-product from the pullet-hatching industry. The bird will weigh about 7½ pounds and have consumed some 35 pounds of food.

Table duckling production

As has already been pointed out in the chapter dealing with table ducklings, the market is limited and only those producers who have a guaranteed outlet should consider this type of meat production.

The following figures may be used as a guide for table ducklings between eight to nine weeks old and weighing approximately 6 pounds liveweight.

	New pence
Day-old duckling	15.0
Food	40.0
Labour	8.0
Heat and light	0.8
Depreciation	1.5
Mortality	1.5
	66.8

Turkey production

The following figures relate to a large type of turkey marketed at twenty-four weeks of age:

	New pence
Day-old poult	35.0
Food	125.0
Labour	12.5
Mortality	7.5
Sundries	10.0
	190.0

No mention has been made of 'returns'. This is because market prices vary from area to area and season to season. The reader should consult his weekly poultry journal for up-to-date market prices of table poultry.

Egg prices have to be calculated on an average of returns over a twelve-monthly period. As the returns for different egg sizes vary, an average return has not been attempted. The reader should follow the weekly prices given by the British Egg Marketing Board to obtain an up-to-date picture.

Index